Saira Moinuddin-Huber

The Leverage Effect of Strategic Behavioral Dimensions in Multi-Channel Negotiations

Verlag Dr. Kovač

Hamburg
2024

VERLAG DR. KOVAČ GMBH
FACHVERLAG FÜR WISSENSCHAFTLICHE LITERATUR

Leverkusenstr. 13 · 22761 Hamburg · Tel. 040 - 39 88 80-0 · Fax 040 - 39 88 80-55

E-Mail info@verlagdrkovac.de · Internet www.verlagdrkovac.de

D100

Bibliografische Information der Deutschen Nationalbibliothek
Die Deutsche Nationalbibliothek verzeichnet diese Publikation
in der Deutschen Nationalbibliografie;
detaillierte bibliografische Daten sind im Internet
über http://dnb.d-nb.de abrufbar.

ISSN: 2365-7898 (Print)

ISBN: 978-3-339-13976-4
eISBN: 978-3-339-13977-1

Zugl.: Dissertation, Universität Hohenheim, 2024

Umschlagfoto: © Karla Fritze

Printed in Germany

Gedruckt auf holz-, chlor- und säurefreiem, alterungsbeständigem Papier. Archivbeständig nach ANSI 3948 und ISO 9706.

Vorwort zur Schriftenreihe

Verhandlungen spielen in Wirtschaft und Gesellschaft, aber auch im Alltagsleben eine entscheidende Rolle. Im Grunde sind die meisten Entscheidungen, an denen mehr als eine Person oder Partei beteiligt sind, das Ergebnis mehr oder weniger aufwendiger Verhandlungsprozesse.

Angesichts der Omnipräsenz von Verhandlungen in praktisch allen Lebensbereichen verwundert es nicht, dass sich viele wissenschaftliche Einzeldisziplinen mit dem Thema „Verhandlungen" aus ihrer Perspektive auseinandersetzen. Zu nennen sind an dieser Stelle beispielsweise die Psychologie, die Politologie, die Rechtswissenschaft, die Informatik, die Soziologie, die Betriebswirtschaftslehre, die Volkswirtschaftslehre oder die Mathematik.

Auch wenn diese wissenschaftliche Multidisziplinarität auf der einen Seite eine breite Ausleuchtung des Themenbereichs Verhandlungen sicherstellt, hat diese auf der anderen Seite auch dazu geführt, dass sich innerhalb der Verhandlungsforschung viele isolierte Forschungslinien entwickelt haben, die kaum miteinander verwoben sind. Die Folge sind parallele Forschungsbemühungen zu den gleichen Forschungsfragen oder widersprüchliche Forschungsergebnisse, die nicht selten darauf beruhen, dass vorliegende Forschungserkenntnisse anderer Teildisziplinen zum gleichen Thema nicht beachtet werden.

Um mehr Forschungssynergien zu ermöglichen, bedarf es einer Intensivierung des wissenschaftlichen Austausches innerhalb der Verhandlungsforschung, aber auch zwischen Verhandlungsforschung und Verhandlungspraxis. Zu diesem Zweck wurde 2013 die Negotiation Academy Potsdam (www.negotiation-academy-potsdam.de) an der Universität Potsdam gegründet. Diese erste universitäre Verhandlungsakademie verfolgt das Ziel des Wissenschafts- und des Wissenschafts-Praxis-Austausches im Bereich der Verhandlungsforschung. Leitbild der Aktivitäten der Akademie ist dabei ein ganzheitliches Verständnis von Verhandlungen als Managementprozess. Betrachtet werden damit einhergehend neben der eigentlichen Verhandlungsführung vor allem auch vor- und nachgelagerte Managementaufgaben bei Verhandlungen. Verhandlungen werden so als Management-Prozess verstanden, der aus den einzelnen Prozessschritten Verhandlungsanalyse („Welche Informationen werden für Verhandlungen benötigt?"), Verhandlungsorganisation („Wie sollte mit und in Teams verhandelt werden?"), Verhandlungsvorbereitung („Wie können Ziele, Strategien und Taktiken festgelegt werden?"), Verhandlungsführung („Wie können Verhandlungen erfolgreich gesteuert werden?") und Verhandlungscontrolling („Wie kann man erfolgreiche von nicht-erfolgreichen Verhandlungen unterscheiden und wie kann ein Verbesserungsprozess für zukünftige Verhandlungen angestoßen werden?") besteht.

Die vorliegende Schriftenreihe zum Verhandlungsmanagement soll auf wissenschaftliche Arbeiten aufmerksam machen, die diesem Forschungsleitbild folgen und Forschungsfragen im Themenfeld „Verhandlungen" aus einer Managementperspektive untersuchen. Hierdurch soll der wissenschaftliche Diskurs im Bereich der betriebswirtschaftlichen Verhandlungsforschung gefördert und den in dieser Schriftenreihe aufgenommenen Arbeiten eine größere Sichtbarkeit verschafft werden.

Potsdam/Hohenheim, im Herbst 2015

Prof. Dr. Uta Herbst

Prof. Dr. Markus Voeth

Geleitwort

Globalisierung und Digitalisierung führen in Unternehmen zu komplexen Verhandlungssituationen, die oftmals durch Mehrparteien-Interaktionen, interkulturelle Beziehungen zwischen den Verhandlungsparteien oder vielschichtige Verhandlungsgegenstände gekennzeichnet sind. Der strategische und taktische Einsatz von Kommunikationsmedien stellt dabei in solchen Verhandlungskontexten einen wichtigen Erfolgsfaktor und einigungsfördernden Aspekt dar. Insbesondere kann dabei der parallele Einsatz verschiedener Verhandlungsmedien einen Stellhebel für eine Verbesserung der Verhandlungsleistung darstellen.

Das Themengebiet „Multi-Channel Verhandlung“ wurde jedoch bislang – trotz einer relativ großen Anzahl an Untersuchungen zum Einsatz von Verhandlungsmedien – noch nicht umfassend wissenschaftlich ergründet. Daher wendet sich die vorliegende Arbeit, die sich mit dem Gebiet der Multi-Channel Verhandlungen beschäftigt, einem wichtigen, zugleich jedoch auch noch wenig erforschtem Gebiet zu. Das Ziel der Forschungsarbeit ist es dabei, den strategischen Einsatz von Mehrkanalverhandlungen zu ergründen.

Insgesamt legt die Verfasserin eine interessante und insgesamt ansprechende Arbeit zu dieser aktuellen, zugleich aber auch anspruchsvollen Fragestellung vor. In der von der Verfasserin durchgeführten Studie kommt sie nicht nur zu einigen entscheidenden Ergebnissen, sondern zeigt auch ihre Fähigkeit zu empirischer Forschung auf. Angesichts des imposanten internationalen Datensatzes sowie der grundlegenden Erkenntnisse, zu der die Verfasserin in ihrer Arbeit gelangt, wünsche ich der Arbeit große Aufmerksamkeit in der internationalen Scientific Community.

Hohenheim, im Februar 2024

Prof. Dr. Markus Voeth

Danksagung

Im Februar 2024 habe ich mit dieser Monografie meine Promotion an der Fakultät für Wirtschafts- und Sozialwissenschaften der Universität Hohenheim erfolgreich abgeschlossen. An dieser Stelle bedanke ich mich für die Unterstützung, die ich während meiner Promotionszeit als externe Doktorandin des Lehrstuhls für Marketing & Business Development auf unterschiedliche Weise erfahren habe und was zum Gelingen dieses Werkes beigetragen hat.

Allen voran gilt mein besonderer Dank meinem Doktorvater Prof. Dr. Markus Voeth. Er hat mich bei der Wahl und der inhaltlichen Ausgestaltung meines Promotionsthemas unterstützt. Für sein entgegengebrachtes Vertrauen in meine akademischen Fähigkeiten möchte ich mich von Herzen bedanken. Weiterhin gebührt mein Dank Herrn Prof. Dr. Karsten Hadwich für die Erstellung des Zweitgutachtens und Herrn Prof. Dr. Christian Ernst für die Übernahme des Prüfungsvorsitzes bei meiner Disputation.

Herzlich bedanken möchte ich mich auch bei allen aktuellen und ehemaligen (wissenschaftlichen) Mitarbeitern des Lehrstuhls. Mein besonderer Dank gilt Monika Fielk, Dr. Sandra Haggenmüller und Dr. Michael Oryl.

Frau Prof. Dr. Uta Herbst und ihrem Lehrstuhlteam in Potsdam danke ich für die konstruktiven Diskussionen während der Doktorandenseminare.

Meinen Eltern, Dr. Hasan und Jolanta Moinuddin, gebührt meine tiefe Dankbarkeit für ihre an mich gerichtete stetige Ermutigung zu promovieren und ihre uneingeschränkte Unterstützung während meines Promotionsvorhabens. Ihr Glaube in meine Fähigkeiten war eine treibende Kraft, um dieses Ziel zu erreichen.

Ganz besonders gilt mein Dank – der Liebe meines Lebens – meinem Ehemann Andreas Karl Hans Huber. Mit viel Rücksicht, Humor und Liebe hat er dazu beigetragen, dass ich diese Promotion erfolgreich beenden konnte.

Bad Teinach-Zavelstein, im Februar 2024

Saira Moinuddin-Huber

Table of Contents

List of Tables

List of Figures

1 Introduction

"In pursuit of knowledge, every day something is acquired. In pursuit of wisdom, every day something is dropped."

Lau Tzu, 6th century BC

Lau Tzu – the Chinese philosopher and founder of Taoism in the 6th Century BC – advocated simplicity (Chan, 2015), which in the half-time of this research phase was fundamentally applied to this research to cut through the vastness of multi-channel complexity and focus on root causes for practical success in negotiations. In the process of working on the topic of *complex* multi-channel negotiations, one not only grows older and matures cognitively yet also grows more knowledgeable in understanding. Fittingly, Adair (2005, p. 51) states, "wisdom always tends towards simplicity." As such, leading and steering this work into the harbor of substantial multi-channel conclusions entailed dropping massive amounts of acquired knowledge and, for the pursuit of wisdom, solely drilling into behavioral success factors in channel negotiations. Accordingly, this work acts as a value creation for practical negotiator success by providing insights for advancement in multi-channel negotiations to global negotiators and the international negotiation sciences communities.

This expedition into multi-channel behavioral dimensions embarked into uncertainty with an apparent *performance risk* whether empirical analysis will - after all invested years, resources, and efforts - yield viable results. A strong self-manifestation to captivate countless negotiators daily to participate in this study, paired with a *self-effective* plowing through data, produced statistical results for this work, which speak for themselves. With this, returning to the opening statement of reducing complexity, contrary to expectations, empirical findings of this research reveal the positive effects of complexity in a multi-channel setting, which result in higher effectiveness and efficiency levels among global negotiators. A review by Koch & Shepperd (2004) even identified self-complexity leads to successful adaptation in adverse situations (Dixon & Baumeister, 1991; Linville, 1985, 1987). This identification brings the research voyage back to the zero point of the Socratic Paradox that "true knowledge exists in knowing that you know nothing" (Plato translated by Fowler, 1966) and implies the necessity for uninterrupted research on the multiple facets of media in negotiations.

Media Promises Success on the Quest for Performance

Faced with incessant evolving global dynamics, companies relentlessly seek guidance on enhancing their negotiation performance (Ertel, 1999; Fisher et al., 2011; Lewicki & Hiam, 2011; Mnookin et al., 2017; Shell, 2006; Thompson, 2011; Voeth et al., 2018), often through negotiator skill training (Fischer & Fisher-Yoshida, 2017). Organizational negotiations guarantee sustainable business development and consequently necessitate a management of these negotiations (Caputo et al., 2019). Correspondingly, a recent study conducted by Voeth et al. (2022) confirms that 70% of the participating German managers require negotiation support, especially during the initiation phase. Even on an international level, complex global economic relations demand managers to be engaged in planning, executing, and following up on negotiations (O'Brien et al., 2008). Managers constantly search for measures to improve their negotiation results (Diamond & Weiss, 2013; Lewicki & Litterer, 1985; Raiffa et al., 2002; Shell, 2006; Watkins, 2014). In essence, negotiating effectively "requires developing a complex, nuanced, and broadly applicable understanding of negotiation" (Loewenstein & Thompson, 2006, p. 92), and it entails learning how to negotiate.

A negotiation is a dyadic exchange with partially different preferences over one or more objects (Bazerman & Moore, 2013; Henderson, 2011; Lewicki et al., 2015; Voeth et al., 2015). The holistic nature of a negotiation applies to the private (Bazerman et al., 1992; Thompson, 2005), political (Bazerman et al., 1992; Curtin & Miller, 2011; Donohue et al., 1991; Faizullaev, 2014), legal (Hollander-Blumoff, 2017) and business context as "all exchanges, especially international agreements, include some elements of negotiation" (Mayfield et al., 1998, pp. 21-24). Characteristics of negotiation are its cooperative and competitive elements (Galinsky & Schweitzer, 2015; Halevy & Phillips, 2015; Pruitt & Carnevale, 1993; Yip & Schweinsberg, 2017). In this agreement progression, in which negotiators attempt to find a solution according to their preference (Raiffa, 1982), a targeted use of negotiation media promises success (Adair & Loewenstein, 2013; De Dreu & Van Kleef, 2004a, 2004b; Dwyer, 2018; Salacuse, 2005; Veiga de Paula et al., 2004) and effective communication is a crucial driver for it (Thompson & DeHarpport, 1998). In consequence, media channels - as a contextual factor - determine the frame and are vehicles for the communication process, affecting intercommunication and thus impacting negotiation effectiveness (Yates & Orlikowski, 1992). On the social-psychological side, negotiation channels reveal behavioral insights by enabling access to social cues and providing influence and support to the individual negotiation style (Fulk et al., 1990; Purdy et al., 2000).

An Evolving Media Landscape

It is external factors and the level of digitization in business that transform negotiations, and with this, new forms and channels of negotiation emerge (Lax & Sebenius, 1986; Weiss, 2006; Zartman, 1994, 2003), thus increasing speed, novelty, interconnectedness, and complexity (Ebner, 2017; Macharzina & Wolf, 2023). Digital technologies are an integral part of the processes of contemporary industrial entities (Eidhoff, 2017), leading to a digital daily routine as the new normality for global negotiators both professionally and in private (Danaher & Rossiter, 2011), working under augmented complexity and uncertainty (Haggenmüller et al., 2023). Media in negotiation can be divided into classic and new (Galin et al., 2007; Poole et al., 1992; Purdy et al., 2000). Negotiations conducted without a computer, or the Internet are classified as analog or traditional media, for instance, face-to-face or phone calls (Danaher & Rossiter, 2011; Voeth & Herbst, 2015). Negotiations managed via the computer or Internet are categorized as digital and fall under new negotiation channels (Schoop, 2021b). Modern digital communication channels in negotiations range from social media to collaboration tools such as Microsoft Teams, Google, Skype, and Zoom. In today's setting, social media is an essential modern digital communication channel in negotiations (Ashbrook & Zalba, 2021) with well-accepted social media platforms acting as digital negotiation canals such as Facebook, Twitter, and LinkedIn (Low & Anshari, 2013; Seyff et al., 2015) for broadcasting diverse multi-media content (Jung & Moro, 2014; Kietzmann et al., 2011; Motameni & Nordstrom, 2014). A new normal is instant messaging with prominent digital applications such as WhatsApp (Baron, 2013; Johnson & Cooper, 2015; Kessler, 2008) offering a unique selling proposition of unregulated, real-time sharing of information. Additionally, FaceTime and Skype are video calling apps that permit users to voice and video call remotely. Collectively, researchers assume that the progression of digital media usurps conventional media (Hayles, 2012), which is a valid assumption to evaluate within the scope of this study.

Perspectives on Media Choice Introduce Multi-Channel Negotiations

A medium has objective physical characteristics – such as the transmission of voice or documentation - referred to as media capabilities to support successful communication (Dennis & Robert, 2009). These multifold capabilities are often beyond comprehension for communication participants and thus often not used appropriately (DeSanctis & Poole, 1994). Media capabilities are observed and rated subjectively between individuals or by the same individual in different instances (Carlson & Zmud, 1999; Fulk et al., 1990; Walther, 1992), and channel characteristics primarily determine channel choice (Macik et al., 2012; Verhoef et al., 2007). The comprehension of user assessment of media capabilities and its consequential impact on media selection is vital in a multi-channel environment (Maity & Dass, 2014; Pookulangara et al., 2011; Wang et al., 2016; Yu et al., 2011). According to Suh (1999) and Fulk et al. (1990), properties of media lead to individual observations of these channel features, and individual rationality then matches media with task requirements. A "social information perspective" (Suh, 1999, p. 299) of channel choice is subjectively induced by positions and statements of others, an assessment of coworkers, and rationally decided upon past statements, behaviors, and social norms making channel attributes the link between negotiators (Danaher & Rossiter, 2011).

Evidently, the choice of negotiation media today is vast compared to decades ago (Agndal et al., 2017) as the only boundary condition which has changed and evolved the most over time (Geiger, 2020). Abundant studies conclude that communication media has an essential effect on negotiations (Poole et al., 1992). Thereupon, an essential requirement to comprehend the effects of communication media on negotiations is more present than ever as the digital transformation offers negotiators a variety of media choices, making it imperative to be cautious in selecting negotiation channels (Purdy et al., 2000).

Lasswells' Communication Theory (1948, p. 216) is fundamental to unpuzzling media choice by asking: "Who says what to whom in what channel with what effect?" In this context, interdisciplinary research denotes multi-channel terminology as a sequenced or parallel use of multiple communication media to send a message effectively and efficiently to a target group (Kunz & Kunz, 2014a, 2014b). Subsequently, after decades of negotiation research, the most prominent unknown territory within the sphere of negotiation media is the mixture of negotiation channels (Ambrose et al., 2008; Damian et al., 2008; Dennis et al., 2008; Geiger, 2020). Only today, these questions are rarely part of research and imply the need for further investigation (Bazerman et al., 2000; Neale & Northcraft, 1991).

1.1 Research Interest & Relevance

Research on channel coordination as a negotiation topic first emerged in the decade from 2003 to 2012 (Meister, 2013). Despite countless studies on negotiation media utilized during negotiation processes, there is still a need for a clear and consistent theoretical foundation for multi-channel negotiations. Few empirical studies exist regarding the impacts of using multiple media on the negotiation outcome (Polychroniou, 2014). By formulating and investigating the following research question conceptually, the objective is to merge fundamentals spread across different disciplines, create a solid theoretical base, and derive initial conclusions for negotiations:

(1) *What are multi-channel negotiations, and how are these derived from interdisciplinary negotiation research?*

The existence and need for multi-channel negotiations are identified in a systematic literature review covering a publication period of 30 years. This study compiles research on channel use in negotiations over three decades, from 1990 to 2020. The foundation for research on multi-channel negotiation is the conception of Media Theory (McLuhan, 1970). In this concept, McLuhan (1970) declares that the channel entails all meanings of the discourse, which implicates how communication media affect differently on the human sensorium.
A definition of multi-channel communication and marketing goes back to the 1970s, blending different communication, distribution, and promotional channels to maximize marketing potential (Mallen, 1973). Considering this background, interdisciplinary research describes multi-, cross-, or omni-channels as seamless communication via channels (Cummins et al., 2016; Manser Payne et al., 2017; Park & Lee, 2017; Verhoef et al., 2015). Transferring this concept to negotiation translates into modifying the negotiation process through modes of communication media (Putnam & Carcasson, 1997). In accordance, a multi-media approach to support negotiations is recommended (Putnam & Roloff, 1992).

Through this research study, a multi-channel negotiation is defined as the employment of more than two communication channels – analog, digital, or both – during the negotiation phases (Ambrose et al., 2008; Arthi, 2009; Barkhi et al., 2004; Damian, 2002, 2003, 2008; Dennis et al., 2008; Gaines et al., 2000; Geiger, 2020; Khan et al., 2011; Suh, 1999; Swaab, 2012; Yuan et al., 2003). A multi-channel negotiation emerges through channels' sequenced or parallel use (Barley, 1991; Greenhalgh & Chapman, 1995).

As expected, using multiple channels makes the negotiation process dynamic and challenging (Putnam & Holmer, 1992). Diverse channels are applied - face-to-face, phone calls, texting, and electronic media - making negotiations inherently multi-media (Bazerman & Moore, 2013). Often electronic replaces traditional negotiation media, transforming the negotiation process caused, among other reasons, by travel and cost advantages (Chiu et al., 2005; MacDonald & Smith, 2004; Poole & DeSanctis, 1990; Shakun, 2005). The choice of channels for negotiators is versatile (Agndal et al., 2017), bargaining parties employ several channels simultaneously (Agndal, 2007) and often prefer one channel (Yuan et al., 2003) in correspondence with the Uses and Gratification Theory by Katz et al. (1973). As in each case, the channel affects the clarity of the information exchange (Sitkin et al., 1992), the most appropriate channel mix is utilized depending on the negotiation (Crump, 2011; Crump & Druckman, 2012).

Strategic communication theories of relevance identified in this systematic literature review apply to a broad communication spectrum and do not center on bargaining only. The collective insights regarding negotiation media use indicate that impacts depend on the context, and no blanket statement is possible; thus it is problematic to derive a successful recipe for multi-channel negotiations from these. Transferring general findings from communication research to a specific negotiation process requires validation to advance negotiation research.

Effectiveness & Efficiency of Behavior in Multi-Channel Negotiations

Since the beginning of research on bargaining, the quest for formulas and control levers, which explain the success or failure of negotiations, has been of utmost importance as negotiations determine economic business performance (Crook & Combs, 2007; Dyer et al., 2018; Lavie, 2007), which creates organizational value (Ganesan, 1993; Graham, 1986; Grennan, 2014). A successful negotiation is characterized by its effectiveness and efficiency – by selecting rightful conduct and due to limited resources conducting it rightfully (Drucker, 1974). However, "views of rational media choice and task performance do not explain performance" (Valacich et al., 1994, p. 18). As negotiations leave an economic imprint (Lewicki et al., 2015), examining the effectiveness and efficiency of relevant media on the negotiation has been nonexistent until now (Danaher & Rossiter, 2011).

To reach a consensus in this environment of complex negotiations, managers must utilize multiple negotiation channels, which aligns with the Integrated Marketing Communication Theory by Kitchen & Schultz (1999). This Integrated

Marketing Communication Theory applied to negotiations emphasizes integrating various communication channels to create a seamless and consistent message to the other party. Axiomatically, this theory implies the gravity of measuring the effectiveness and efficiency of communication strategy on its outcome using data and analysis to improve and optimize the strategy continually. Transferred to negotiations, it provides a strategic framework for developing and implementing - as a global appeal by Mayer & Voeth (2022) - effective and efficient negotiation strategies that are result-oriented & data-driven.

As the quest for individual negotiation success prevails (Anderson & Huang, 2006; Mayer & Voeth, 2022; Thompson et al., 2010) rather than focusing on media characteristics, a closer examination of negotiators using media promises the needful insights (Walther, 1992). In the long run, communication transcends media (Walther, 1992), which contradicts McLuhan's theory (1970) that the channel embodies the content of the transaction. To such a degree, success in a negotiation is attributed, among other aspects, to the heterogeneous behavior of negotiation parties (Albers & Hildebrandt, 2006). Especially an individual negotiator (Bazerman et al., 2000) with effective communication behavior can impact the success of a negotiation.

Meister (2013) depicts that negotiation behavior has been under the lens of negotiation sciences since the 1960s (Mathews et al., 1972; Pennington, 1968), with a focus decade from 1983 to 1992 depicting publications on negotiation behavior often in conjunction with negotiation strategies. Performance, in general, is defined as the "behavior evaluated in terms of its contribution to the goals of the organization" (Johnston & Marshall, 2006, p. 412) making the individual's behavior worth a measurement (Zallocco et al., 2009). On that account, executives are concerned about the effectiveness and efficiency of negotiation behavior (Carnevale & Pruitt, 1992; Gelfand & Dyer, 2010; Neale & Bazerman, 1992). Nonetheless, systematic studies on negotiation behavior in corporations are limited, and scientific analyses often focuses on the impact of singular negotiation instruments or techniques (Voeth et al., 2022). Negotiators are often not aware of positive or negative behavioral effects (Adair et al., 2001; Neale & Bazerman, 1991; Thompson & Hastie, 1990; Tinsley & Brett, 2001; Pruitt & Carnevale, 1993) on a multi-channel negotiation process and outcome. To assess negotiation behavior, Voeth et al. (2015, 2018) conducted multiple studies with negotiators in Germany and by following the approach of Voeth et al. (2015, 2018), this research validates how success in multi-channel bargaining depends on the behavior of negotiators. Consequently, analyzing negotiators using media and examining their behavior and communication is imperative to this investigation. In the end, negotiators decide when and how information is exchanged effectively and efficiently through which negotiation channels with negotiation partners (Ambrose et al., 2008; Arthi, 2009; Dennis et al., 2008; Dennis & Robert, 2009; Lasswell, 1948; McLuhan, 1964; Olekalns et al., 2003a, 2003b; Swaab et al., 2012; Walther, 1992, 1994).

Numerous negotiators favor manifold media, employ these, and even shift in media application (Neslin & Shankar, 2009). Under these circumstances, behavioral effects occur – so-called reactions – in individuals, dyads, or groups as a result of using multiple media (Lowery & DeFleur, 1988; Perse, 2001; Tsfati, 2011). Conclusively, multiple media usage impacts behavior and changes behavior (Wangenheim & Bayon, 2014). These findings, transferred to the context of negotiations, indicate that negotiation behavior becomes more complicated with augmented media use. Existing research frequently centers on negotiation behavior in solitary channels and in isolation from alternative media (Yu et al., 2011). Moreover, deviations in findings occur on negotiation behavior in relation to a specific negotiation channel. As a result, negotiation parties' behavior is analyzed using multiple communication channels. For this, latent constructs can aid in the revelation of behavioral effects on the success of an interaction. The measurement of behavior and performance helps to identify a specific behavior as being responsible for an effective and efficient negotiation outcome. In this study, it is assumed that negotiation channels determine negotiator behavior. Opposed to this assumption, several authors report no differences in behavior in general during the use of various communication media (Delaney et al., 1997; Staples & Zhao, 2006; Zornoza et al., 2002). For negotiations specifically, no disparity in behavior using various media channels has been identified (Griffith & Northcraft, 1994; Yang, 2012). This study hypothesizes - opposed to prevalent scientific findings - that multi-channel impact negotiation behavior and challenges ubiquitous systematic rulings in a variable interaction analysis by validating corresponding outcomes with existing research.

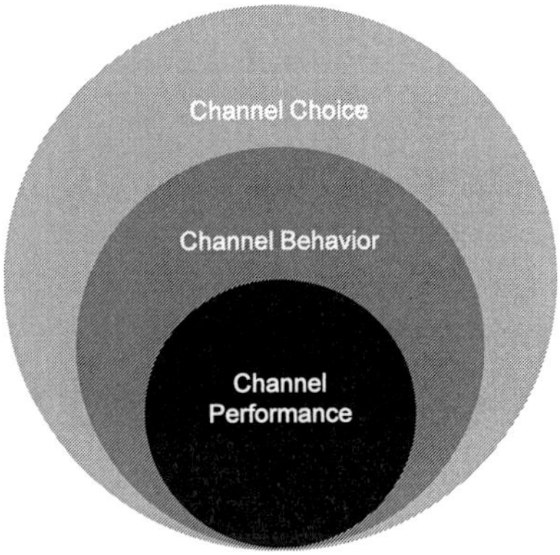

Figure 1. A Three Step Approach to Unlocking Channel Performance

To this day, most negotiation science remains within the boundaries of classic exploration agendas (Agndal et al., 2017; Knöpfle, 2019) and to match the demands of global negotiators for strategic directions, studies of actual negotiations on unknown terrain are imperative (Fells et al., 2015). As no study exists on multi-channel negotiations, this exploration examines how multi-channel negotiations can be strategically employed.

In an ongoing digital transformation of the negotiation media landscape, a deficiency of comprehension prevails regarding these decisive issues. Constructed on negotiation literature, a descriptive behavioral sciences approach, as opposed to the analytical prescriptive approach, is selected to assess negotiation behavior in real-life (Voeth & Herbst, 2015). Complementary to these theoretical findings, the following research question explores empirically:

(2) *How does a specific negotiation behavior impact the effectiveness and efficiency of multi-channel negotiations?*

Grounded in the systematic literature review, an abstract construct (Rossiter, 2008) on multi-channel negotiations aids in understanding higher-level behavioral negotiation strategies. Specifically, a latent construct can contribute to analyzing unobservable behavioral drivers (Coltman & Devinney, 2007) for negotiation success. This latent construct makes understanding the interrelation between latent behavioral constructs and negotiation success possible. A conceptual model for this research study attempts to fulfill the requirements of real-life complexity (Kersten & Lai, 2007; Schoop, 2010; Watkins, 1999) in multi-channel negotiations. To this extent, behavioral dimensions in multi-channel negotiations are explored with a conceptual framework to understand their impact on negotiation performance.

This study uses data from negotiation experts in companies who evaluate behavioral success factors in a multi-channel environment. These behavioral success factors are specified in a questionnaire using measurement models with multiple items (Bergkvist & Rossiter, 2007; Churchill, 1979) and this approach increases the ambition to search for specific behavioral patterns in a negotiation (Thompson, 1990). The application of such a scientific method aims to mature into a clarifying structure that assists practitioners and academics in comprehending heterogenous interrelations in behavior during bargaining phases. A resulting exhibit of interaction disparities tackles a significant challenge in negotiation research. This is to create a framework for exploring modes of bargaining behavior, to attain acceptable outcomes, and to comprehensively understand the interrelations in behavior.

The substance of this paradigm is theoretically deducted by dividing the construct of behavior in multi-channel negotiations into dimensions (Albers & Eggert, 1988). After the definition and comprehension of the causal relationships in this construct, several assumptions are formulated on how this theory is influenced positively or negatively in its outcome. Under these circumstances, behavioral effects in multi-channel negotiations are assessed influencing negotiation effectiveness and efficiency through four behavioral dimensions and three moderator variables from the theoretical assessment. In accordance, the research objective of this analysis is to identify the effect mechanism of behavioral dimensions – specifically self-efficacy, agility, synchronicity, and behavioral control – in multi-channel negotiations, to evaluate data in a structural equation model, then to compare results of the analysis with hypotheses and to subtract measuring error. As a result, a theoretical deduction for the content of constructs of negotiation behavior is derived from negotiation literature (Albers & Eggert, 1988; Voeth & Herbst, 2015). Thereby, an inductive approach is anticipated to form a behavioral multi-channel negotiation theory. Accordingly, the study tests the perceived effectiveness and efficiency of negotiation messages delivered by a specific negotiation behavior through various channels.

Statistical analysis tests the postulated hypotheses of this study for plausibility, and knowledge outside of statistics is also required (Backhaus et al., 2015). According to the Hempel Oppenheim Model by Hempel & Oppenheim (1948), the deductive nomological approach tests hypotheses, which are grounded in theoretical, factual, and logical considerations. It is generally assumed that behavioral dimensions can positively impact the negotiation outcome. Beyond doubt, all behavioral dimensions could also negatively impact negotiation effectiveness and efficiency, yet for reasons of simplicity, no additional hypotheses cover this negative scenario.

As a first behavioral dimension, the applied self-efficacy theory in this research derived from Bandura (1997) and Pekkala & Van Zoonen (2022), reflects the confidence in the ability to evaluate negotiation channels, select these independently, and implement these effectively and efficiently. It is assumed that self-efficacy positively impacts negotiation effectiveness and efficiency.

The second behavioral dimension, agility in a negotiation, is "the capacity to react quickly to rapidly changing circumstances" (Brown & Agnew, 1982, p. 29), which demands concentration on negotiation objectives and matching resources to the demands of changing negotiation circumstances. As follows, agility refers to a negotiator's ability to accelerate activities on critical paths. Derived from agility research of Kumar & Motwani (1995), it directly indicates a negotiator's time-based effectiveness in the negotiation. It is hypothesized that agility is a result of multi-channel negotiations, which impacts negotiation effectiveness and efficiency.

The third behavioral dimension, synchronicity – caused by using diverse negotiation channels – is a coordinated, synchronous behavior of the negotiation parties with a shared goal to create a win-win situation, which can lead to an integrative negotiation (Arthi, 2009; Dennis et al., 2008; Dennis & Robert, 2009). The construct synchronicity contemplates if a multi-channel interaction's information or semantic content in a negotiation process produces a mutual negotiation conception to obtain a sustainable negotiation deliverable. The efficacious matching of negotiation channels with the content determines whether synchronicity is established, thus positively impacting negotiation effectiveness and efficiency.

The final behavioral dimension of behavioral control is grounded in work of Lim (2003). Behavioral control within this study is defined as the negotiator feeling control over objects and processes of the negotiation while using multiple channels. Negotiators prefer media channels over which they believe they have more control instead of channels where perceived control is less likely to positively impact negotiation effectiveness and efficiency.

Supplementary to the behavioral dimensions in the bargaining context, it is imperative to consider moderating effects in complex causal structures (Eggert et al., 2005) to assess under which circumstances a causal relationship is especially strong or distinctly weak. Within the scope of this research, it is assumed that the causal relationship is fragile under certain circumstances. Moderator variables, which impact the main effects negatively and weaken these, are complexity, cognitive resource demands, and performance risk. These moderating effects are not part of the causal relationship and yet influence the impact path and strength between independent and dependent variable relationships (Baron & Kenny, 1986; Sheer, 2017). Anticipated correlations among variables modify if the moderator variable changes. It is assumed that postulated relationships between behavioral dimensions and the negotiation outcome depend on the values of third variables (Ramayah et al., 2018). A priori, it is hypothesized that moderating variables negatively affect the direction and intensity of all construct relationships in this model (Dawson, 2014). Due to reasons of simplicity, a positive impact of moderating effects on the variable relationships has not been constructed.

Intercultural Negotiation Perspectives

Numerous international studies on negotiation practice (Adair & Brett, 2005; Liu et al., 2010; Lügger et al., 2015) emphasize the importance of not only relying on intracultural negotiation perspectives, which are negotiations between national counterparts (Adler et al., 1987; Metcalf et al., 2006). The importance of international negotiations is rarely mirrored in negotiation sciences (Herbst et al., 2011), and negotiation research often depicts optimizing negotiation results while neglecting the negotiation process. For this reason, 733 international negotiation managers evaluate the effectiveness and efficiency of negotiation behavior in multi-channel negotiations. Broadening empirical research to an international setting (Liu et al., 2012; Salmon et al., 2013) follows the assumption that intracultural negotiation cannot be identically transferred to intercultural negotiations. An international sample size provides necessary insights into heterogeneous cultures and corresponding diverse negotiation behavior (Brett, 2014; Graham et al., 1994). It is argued that cultural discrepancies lead to augmented complexity. However, the negotiator's cultivation of an organizational culture implies a similar global negotiation approach (Kaspzark, 2017). The results of this study reveal both cultural discrepancies regarding negotiation channel use and homogenous negotiation behavioral impacts determined by organizational size and culture globally.

Multi-Group Analysis on Channel Quantity

To complete a managerial view on negotiation channel management, this study tests whether a certain quantity of channels is more suitable than others for achieving optimal negotiation results. This method appraises the effectiveness and efficiency of behavior in multi-channel negotiations in terms of how these are evaluated by negotiators and thereby, to assess which negotiation behavior and quantity of channels positively affect the negotiation outcomes. Accordingly, the structural equation model is tested employing a multi-group analysis for comparing sample subgroups with divergent channel quantities. Mirrored through the feedback from global negotiation practitioners, negotiation experts are keen to distinguish the impact on a negotiation if the number of channels employed augments. To identify differences in results, a multi-group analysis examines the impact of channel quantity on the negotiation behavior and outcome. For each of the above groups, an exact estimation of structural parameters applies (Huber et al., 2007). The following section presents the structure and scope of this study.

1.2 Structure & Scope of the Study

This multi-channel negotiation research is structured in eight chapters, as depicted in Figure 2. To answer the research questions comprehensively, the applied methodology of this investigation is divided into two parts 1) a systematic literature review paired with 2) an expansive and in-depth empirical analysis. The database for the broad empirical analysis consists of experienced negotiators, as multi-channel negotiations are of foremost relevance to professional bargaining. Analyzing negotiation behavior in multi-channel negotiations is based on negotiation experts' experiences and assessments collected through an online survey to generate primary data. Correspondingly, Chapter 1 comprises the overview stating the analysis's relevance, the guiding research questions, and the goals, as well as closing the chapter with the overall structure and scope of the study.

A systematic literature review in Chapter 2 identifies relevant research on multi-channel negotiations, applicable strategic communication theories, and essential social-psychological perspectives. Existing research on media use in negotiations, as secondary data, is analyzed to provide an unbiased and comprehensive summary of the present understanding of this topic. By defining the concept of multi-channel negotiations, the first research question is answered, while at the same time identifying further research gaps on this topic, which lead to the imperative second research question. The discoveries from the qualitative literature review are processed to examine the impact of multi-channel in business negotiations, and from this, a concept for a quantitative measurement of multi-channel negotiations is extracted. This research foundation aids in defining behavioral dimensions of self-efficacy, agility, synchronicity, behavioral control, and the moderating effects of complexity, cognitive resource demands, and performance risk in the specific context of multi-channel negotiations. From this, a conceptual framework for the effect mechanism of the specified behavioral dimensions and corresponding moderating effects in multi-channel negotiations is outlined. Grounded in the literature review, corresponding theories are derived, and a quantitative data analysis is applied to validate qualitative research findings in negotiations.

Chapter 3 elaborates on the applied methodology and corresponding data analysis to quantify the effects of behavioral dimensions and moderating variables on outcome effectiveness and efficiency in multi-channel negotiations. Selected indicators are formulated according to success factor studies in the business context, according to Albers & Hildebrandt (2006). A quantitative study is conducted to collect data on the behavioral effect mechanism in multi-channel negotiations. In general, the questionnaire focuses on the employment of media channels in business-to-business negotiations. This quantitative analysis divides into binary segments (see Figure 2). The primary segment of the survey examines negotiators' use of media during the negotiation process, which comprises insight into the type, quantity, and sequence of channels used.

Additionally, preferences for channels and targeted use of media are examined. In a secondary segment, a causal analysis reveals impacts of the specified relationships between variables in multi-channel negotiations. The strategic quality of this work is to contrast behavioral effects during media application on negotiation effectiveness and efficiency. Negotiation channels in the scope of this study are analog channels - phone calls & face-to-face - and digital channels - email, communication platforms, and social media.

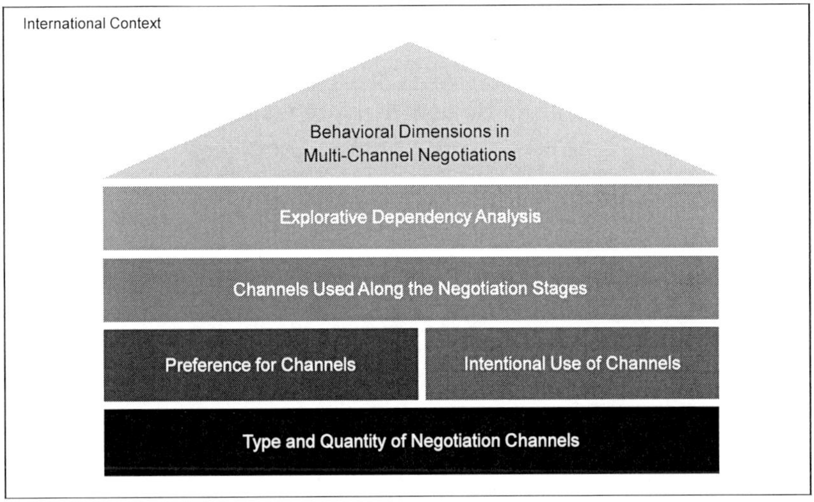

Figure 2. The House of Quantitative Analysis on Channel Negotiations

Finally, the impacts of behavioral dimensions on efficiency and effectiveness in multi-channel negotiations are examined. The findings of this quantitative study postulate a foundation for profound strategic planning and judgments in business negotiations and enhance a prospective development of buyer-seller bargaining. In conclusion, this quantitative study reveals a direction of strategic behavior in multi-channel negotiations. The results are disclosed in Chapter 4 and discussed in Chapter 5. Finally, in Chapter 6, key findings are summarized, and in Chapter 7, the propositions feature potential managerial strategies to administer negotiations more effectively and efficiently and highlight theoretical implications for negotiation sciences. Limitations of this study and the sourcing of prospective research conclude this work in Chapter 8.

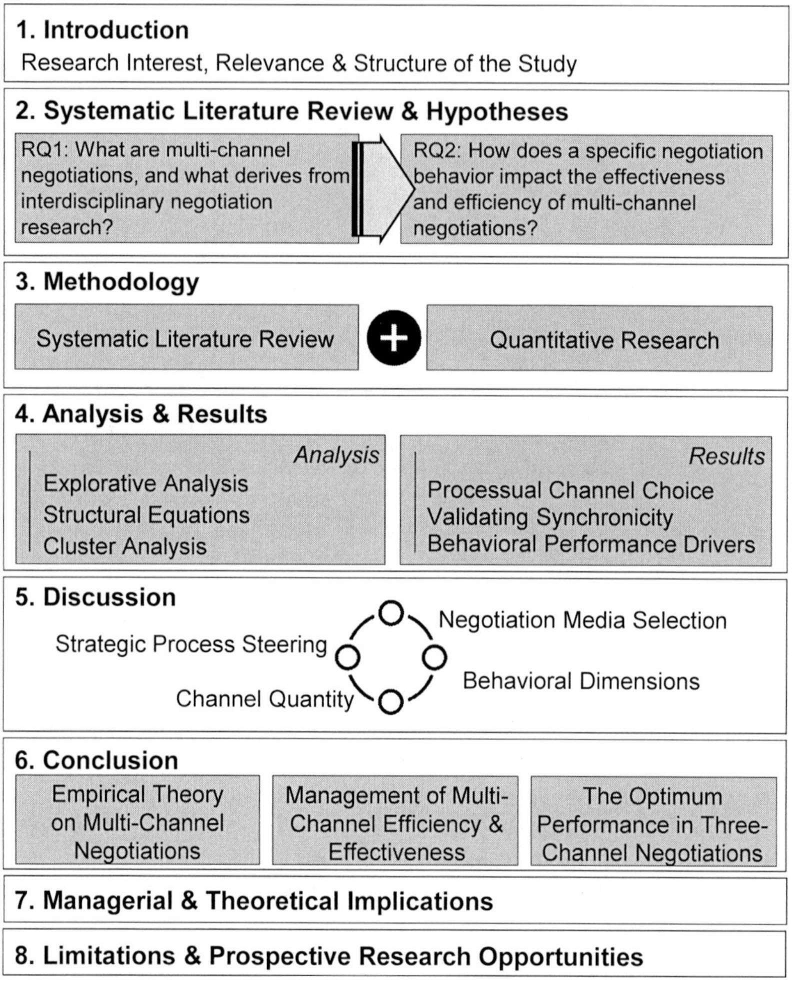

Figure 3. Structure of the Monograph

2 Systematic Literature Review & Hypotheses

This chapter discusses results of the systematic literature review. A multi-channel negotiation is defined, and relevant theories are examined. The general effects of behavior on the course of the negotiation and outcome are analyzed, and reasonable behavioral dimensions generated. Accordingly, hypotheses are derived, and a conceptual framework is developed.

2.1 Media in Negotiations

In essence, negotiations are an arrangement of statements to gain a consensus on divergent interests between two or more parties (Pruitt, 1998; Shin, 2014). On this basis, negotiating is a daily interaction to maintain rapport across layers of hierarchy (Lewicki & Litterer, 1985). This study focuses on the essential daily business negotiations in which a meticulous preparation of the negotiation process promises success (Agndal et al., 2017; Bazerman & Lewicki, 1983; Fang, 2006; Ghauri & Usunier, 2003). Conducting internal, and external negotiations is a daily challenge for managers (Beersma & De Dreu, 2002; Lax & Sebenius, 1986; Herbst et al., 2011; Kulik & Olekalns, 2012; Voeth & Herbst, 2015). Subsequently, effective communication is a crucial driver for negotiation success (Thompson & DeHarpport, 1998). Abundant studies conclude that communication media has an essential effect on negotiations (Poole et al., 1992). In consequence, media channels - as a contextual factor - determine the frame for the communication process, affect intercommunication and thus impact negotiation effectiveness (Yates & Orlikowski, 1992). Moreover, negotiation channels enable access to social cues and provide influence and support to the individual negotiation style (Fulk et al., 1990; Purdy et al. 2000). An essential requirement to comprehend the effects of communication media on negotiations is more present than ever as the digital transformation offers negotiators a variety of media choices, making it imperative to be cautious in selecting negotiation channels (Purdy et al., 2000).

Media in negotiation is divided into classic and new (Galin et al., 2007; Poole et al., 1992; Purdy et al., 2000). Negotiations conducted without a computer, or the Internet are classified as analog or traditional media, for instance face-to-face or phone calls in negotiations (Danaher & Rossiter, 2011). Negotiations managed via the computer or Internet are categorized as digital and fall under new negotiation channels (Schoop, 2021b). Modern digital communication channels in negotiations range from social media to collaboration tools such as Microsoft Teams, Google, Skype, and Zoom. Social media is an essential modern digital communication channel in negotiation (Ashbrook & Zalba, 2021). Well accepted social media platforms also acting as digital negotiation canals are Facebook, Twitter, and LinkedIn (Low & Anshari, 2013; Seyff et al., 2015). Diverse multimedia content can be broadcasted via these platforms (Jung & Moro, 2014; Kietzmann et al., 2011; Motameni & Nordstrom, 2014). A new normal is instant messaging with prominent digital applications such as the popular WhatsApp (Baron, 2013; Johnson & Cooper, 2015; Kessler, 2008). The unique selling proposition of these applications are the sharing of information in real-time. Furthermore, FaceTime and Skype are video calling apps, which permit users to voice and video call remotely.

Not only in corporate settings, yet also in private settings, Skype and Zoom apply predominantly for virtual conferencing (Matulin et al., 2021; Maulana, 2023). Cisco and Avaya are examples for amalgamated platforms to orchestrate all corporate interactions (Bolton et al., 2016; Evans, 2004; Reisinger et al., 2022). Collaboration tools like Slack and Microsoft Teams permit operators to mutually perform in assignments simultaneously (Lomas et al., 2008; Marquez et al., 2004). Digital negotiation channels are rapid, mostly informal, and difficult to document. Along these lines, digital negotiation channels are often used with other negotiation channels, such as email or instant messaging (Damian, 2002, 2003, 2008; Yuan, 2003). The study of Danaher & Rossiter (2011) identifies that, on the one hand, email is easier to reject, and it is generally rated more negatively. On that account, negotiators often tend to be overly positive in respect to the overall success rate of email. Findings undermine that email is more trustworthy. Also, email to most operators is "convenient, acceptable, and appropriate" (Danaher & Rossiter, 2011, p. 18). Besides, email is rapid and effective, and individuals concentrate on the negotiation. Overall, negotiators are more inclined to accept vast volumes of email exchange as opposed to interactions with other media. Primarily, email has some favorable attributes. Nonetheless, email only sometimes translates into more effective negotiation outcomes (Geiger, 2020).

In other words, no single communication technology will likely prove adequate for all task situations or phases of a single task (Mennecke, 2000). Instead, managers should investigate technologies that are easily and rapidly configurable by users to support multiple communication modalities. One technology with the mentioned capabilities is video conferencing, which blends voice, video, and synchronous data sharing. With this environment, users are empowered to choose the most fitting channel configuration for a specific setting (Fish et al.,

1993). Also, the specified technologies allow users to move between lean and rich communication easily and rapidly during the same communication episode (Alavi et al., 1995). This flexibility holds promise for supporting a broader range of organizational communications.

A medium has objective physical characteristics – such as the transmission of voice or documentation - referred to as media capabilities to support successful communication (Dennis & Robert, 2009). These capabilities are often beyond comprehension for communication participants and thus not used appropriately (DeSanctis & Poole, 1994). Furthermore, media capabilities are observed subjectively between individuals or by the same individual in different instances (Carlson & Zmud, 1999; Fulk et al., 1990; Walther, 1992) and channel characteristics primarily determine channel choice (Macik et al., 2012; Verhoef et al., 2007). Comprehension of user assessment of media capabilities and its consequential impact on media selection is vital in a multi-channel environment (Maity & Dass, 2014; Pookulangara et al., 2011; Wang et al., 2016; Yu et al., 2011). According to Suh (1999) derived from Fulk et al. (1990), properties of media lead to individual observations of these channel features and assignment elements. This individual rationality then influences the matching of media with task requirements. A "social information perspective" (Suh, 1999, p. 299) of channel choice is subjectively induced by, for instance, positions and statements of others, an assessment of coworkers, and rationally decided upon past statements, behaviors, and social norms. Channel attributes become the link between negotiators (Danaher & Rossiter, 2011). In the following section, media choice is discussed.

2.1.1 Choosing a Spot-On Channel at an Exact Time

This study agrees that classical "views of rational media choice and task performance do not explain actual performance" (Valacich et al., 1994, p. 18). Media theories must carefully consider the entire group process, not simply task execution activities. The preceding outcomes of this analysis are of importance to bargainers charged with selecting and deploying communication technologies for negotiations within organizations. The study results suggest that issues beyond normative characteristics of tasks and technology influence users' communication performance. It is not the task characteristics that matter as the task characteristics do not influence communication performance.

While it is indispensable "to document the perceived advantages and disadvantages of traditional and new channels, managers are also interested in bottom-line effectiveness" (Danaher & Rossiter, 2011, p. 9). This study follows the path of organizational research by analyzing measures for effective performance creation by Albers & Hildebrandt (2006). By following the path of organizational research media can be compared regarding its effectiveness in negotiations (Purdy et al., 2000). Following this approach, the perceived effectiveness of negotiation messages delivered by different channels can be tested. Lasswell (1948) appropriately defines an interaction process as sending a message through channels to an audience with a corresponding effect. This traditional concept of effectual interactions implies “that the sender's message will be most effective when the sender has correctly anticipated the needs and preferences of the receiver” (Danaher & Rossiter, 2011, p. 9). A message-centered approach for effective communication is proposed by the Shannon-Weaver Model (Shannon, 1949), whereas McLuhan (1964) suggest a focus on media to conclude an interaction effectually. The recommendation to achieve a successful negotiation entails the choice of media channel. Henceforth, the sender should also use the receiver's preferred channel for effective communication. Preceding investigations confirmed that individuals own clear inclinations toward media (Bothorel, 2021; Flanagin & Metzger, 2001; Westmyer et al., 1998). Equally, dispatchers of communicative content evidently favor vehicles of media (Rice et al., 1998). On that account, a divergence-dilemma emerges if negotiators focus on individual channel predilection, thus deviating from channels favored by the opponent (Danaher & Rossiter, 2011). According to Danaher & Rossiter (2011) an assessment reveals if the deviation in channel preference is in the end more harmful to an effectful communicative discourse than harmonizing channel favoritism.

A behavioral intention during a negotiation impacts channel choice and the negotiation outcome (Andaleeb, 1995; Pruitt, 1983; Sheu, 2016). Negotiators assess the type and content of the negotiation to select negotiation channels (Bichler et al., 2003). Hence, the choice of negotiation channels does not happen randomly. Channel choice depends on the behavioral intention of a negotiator, the attitude of the bargaining party towards media channels, and trust of one bargainer in the other negotiation party. Negotiators' often own their media choice. Nevertheless, habitually, operating procedures determine the media use (Trevino et al., 1987). The first step is to define the negotiation objective to negotiate with a spot-on channel at a precise time (Bazerman et al., 2000). This prior objective formulation aids in selecting the most appropriate channel. The second step is to consider the negotiation partner and the corresponding preferences. This focus on the negotiation partner's channel preference supports in determining which media offers the most opportunities for communicative success. The closing stage is to assess and quantify the outcomes of the bargaining phase.

2.1.2 Definition of a Multi-Channel Negotiation

The foundation for research on multi-channel negotiation is the conception of Media Theory (McLuhan, 1970). In the concept of Media Theory McLuhan declares that the channel entails all meaning of the discourse, which means communication media affect differently on the human sensorium. A definition of multi-channel communication and marketing goes back to the 1970s, blending different communication, distribution, and promotional channels to maximize marketing potential (Mallen, 1973). Considering this background, interdisciplinary research describes multi-, cross-, or omni-channels as (seamless) communication via channels (Cummins et al., 2016; Manser Payne et al., 2017; Park & Lee, 2017; Verhoef et al., 2015). Transferring this concept to negotiation translates into modifying the negotiation process through modes of communication media (Putnam & Carcasson, 1997). In accordance, a multi-media approach to support negotiations is recommended (Putnam & Roloff, 1992). Through this research study, a multi-channel negotiation is defined as the employment of more than two communication channels – analog, digital or both – during the negotiation phases (Ambrose et al., 2008; Arthi, 2009; Barkhi et al., 2004; Damian, 2002, 2003, 2008; Dennis et al., 2008; Gaines et al., 2000; Geiger, 2020; Khan et al., 2011; Suh, 1999; Swaab, 2012; Yuan et al., 2003). A multi-channel negotiation emerges through channels' sequenced or parallel use (Barley, 1991; Greenhalgh & Chapman, 1995). Diverse (digital) channels are applied (face-to-face, phone calls, texting, and electronic media), making negotiations inherently multi-media (Bazerman & Moore, 2013). Nowadays, electronic often replaces traditional negotiation media, transforming the negotiation process. Due to travel and cost advantages, countless negotiations are conducted electronically (Chiu et al., 2005; MacDonald & Smith, 2004; Poole & DeSanctis, 1990; Shakun, 2005). The choice of channels for negotiators is versatile (Agndal et al., 2017). It follows that, negotiators use several channels simultaneously (Agndal, 2007) and often prefer one channel (Yuan et al., 2003) in correspondence with the Uses and Gratification Theory by Katz et al. (1973). In each case, the channel affects the clarity of the information exchange (Sitkin et al., 1992), and depending on the negotiation, the most appropriate channel mix is utilized (Crump, 2011; Crump & Druckman, 2012).

Research on multi-channel negotiation emerges for the first time during the highest peak of empirical research on communication media from 2002 to 2016. The analysis of communication media in multi-channel negotiations focuses on the most prominent media used in business negotiations: face-to-face, video, audio, phone calls, messaging, email, and social media. The following researchers analyzed existing theory on negotiation media and developed new theories on multi-channel negotiations: Ambrose et al., 2008; Arthi, 2009; Barkhi et al., 2004; Damian, 2002; Damian et al., 2008; Damian et al., 2003; Dennis et al., 2008; Gaines et al., 2000; Khan et al., 2011; Yuan et al., 2003.

In a purchasing environment, the study of Ambrose et al. (2008) detects a rapport building mechanism to induce channel choice. The corresponding findings validate threefold factors influencing channel choice, namely interaction prerequisites of bargainers, the maturity of affiliation between the opponents, and the procurement context. In essence, limiting the choice of negotiation media would be counterproductive to the negotiation process and outcome.

Table 1. A Media Mix Discourse in Negotiation Research

Author	Media Mix Discourse in Negotiation Literature
Lax & Sebenius, 1986; Weiss, 2006; Zartman, 1994, 2003	The level of digitization in business transforms negotiations and as a result new forms and channels of negotiation emerge.
Agndal et al., 2017	Evidently, the choice of negotiation media today as compared to decades ago is vast.
Bazerman & Moore, 2013	Diverse (digital) channels are applied (face-to-face, phone, text and electronic) making negotiations inherently multi-media.
Putnam & Roloff, 1992	A multi-media approach to support negotiations is recommended.
Sitkin et al., 1992	In each case, the channel affects the clarity of the information exchange.
Crump & Druckman, 2012; Crump, 2011	Depending on the negotiation, the most suitable channel mix is utilized.
Swaab et al., 2012	A recommendation of multi-channel negotiation - audio, visual and synchronicity - to reinforce the positive negotiation effect for neutral negotiators, mitigate non-cooperative negotiation behavior and not to interfere with cooperative negotiators in achieving joint profit.

Similarly, Damian (2002), in software requirement negotiations, enriches a multi-channel computer-mediated negotiation - conducted primarily via audio and video - with a shared electronic workspace in the form of a notepad. Damian (2002) even goes one step further and defines an epoch of multi-channel computer-mediated communication. Damian et al. (2003) discover that a negotiation medium needs room to take effect on the negotiation, and this effect is disturbed by interference with potential social behavior. Closing the research series on multi-channel negotiation, Damian et al. (2008) conclude that combining lean and rich media improves negotiation effectiveness, especially under manifold viewpoints.

Dennis et al. (2008) identify that solitary media is mediocre in effectiveness than multi-channels (Shahriza & Heckman, 2005; Watson-Manheim & Bélanger, 2007). Communication performance improves substantially while parties apply a diversity of channels to accomplish an assignment as opposed to solitary media utilization. This media application to a negotiation task requires an assessment of the device's media capabilities (Bretz, 1983; Heeter et al., 1989; Walther, 1992). For example, a smartphone offers synchronous instant messaging, voice, video, asynchronous email, and text messaging. One device is not a single entity but offers many digital communication media choices with these media capabilities.

Impacts of multi-channel in a bargaining context must be tested considering numerous outcome variables (Yuan et al., 2003). These effects have a potential influence on the observed performance of the overall negotiation. In the subsequent section, this analysis considers relevant strategic communication theories and social-psychological perspectives. This theoretical foundation acts as a base to detail the method applied to respond to the postulated investigation.

2.2 Strategic Communication Theories

To comprehend the concept of multi-channel negotiations, it is necessary to review communication theories of a strategic nature that center around media and assignment' features and advocate a solitary channel or channel arrangement for communicative interactions (Geiger, 2020). In relation to multi-channel negotiations, the following strategic communication theories are assessed: Media Richness Theory (Daft & Lengel, 1986) in combination with the Task and Media Fit Hypothesis (Hollingshead et al., 1993; McGrath, 1993), Uses & Gratification Theory (Katz et al., 1973; Nysveen et al., 2005), the Theoretical Framework of Integrated Marketing Communication by Kitchen and Schultz (1999), Grounding Theory in Communication (Clark & Brennan, 1991; Clark & Schaefer, 1989) and Media Synchronicity Theory (Dennis et al., 2008).

Table 2. Strategic Communication Theories

Strategic Communication Theories	Author
Communication Theory	Lasswell, 1948
The Medium is the Message	McLuhan, 1962
Grounding in Communication	Clark & Brennan, 1991
Framework of Integrated Marketing Communication	Kitchen & Schultz, 1999
Media Synchronicity Theory	Dennis et al., 2008
Uses and Gratification Theory	Katz et al., 1973

Note: This table does not present all applicable and available theory. It is only a selection of relevant theories in relation to the topic of multi-channel negotiations.

2.2.1 Contrasting Media Richness Theory & Task-Media Fit Hypothesis

According to Media Richness Theory by Daft & Lengel (1986), task performance improves as soon as requirements for information equal a channel's richness in information. Channels with a richness transmission capacity are better suited to equivocal tasks and based on studies by Avritzer et al. (2015, p. 25) "less rich media is best suited to tasks with lack of information." Suh (1999) assigns communication media a certain level of richness. In this analysis, the richest media in descending order is face-to-face, phone call and, with only a moderate degree of media richness, scripted correspondence (Bazerman et al., 2000). Media Richness Theory identifies the highest effectual media as opposed to recommendations for professional channel selection (Daft & Lengel, 1986; Dennis & Kinney, 1998; Hartmann, 2009). Despite this approach, knowing the projected effectiveness of media is indeed a cause for channel selection (Daft et al., 1987; Hartmann, 2009). Attempts of transferring the concept of media richness to digital media lack substance (Afflerbach, 2020; Burke & Chidambaram, 1999; Carnevale et al., 1981; Dennis & Kinney, 1998; El-Shinnawy & Markus, 1992).

The activity to negotiate finds itself among an extensive topology of tasks offered by McGrath (1984) for better differentiation to assign media compiled by Valacich et al. (1994, p. 12):

- "generate ideas or plans,
- choose a correct answer or a preferred answer,
- negotiate conflicting views or conflicting interests,
- and execute in contests against another as individuals or group or competition with external performance standards".

Along these lines, dyadic communication builds on task and media fit (Suh, 1999). Input factors for communication are assignment requirements, which can be intellectual in nature or a negotiation, and channel functionalities in the form of script, visuals, or sound. Output factors are the condition for consensus building and decision time, and the output factor of task satisfaction consists of process and outcome satisfaction (Tarafdar et al., 2010). In effect, demands for media richness vary depending on the assignment and context (Kock, 2005). As dependencies augment, decision making among individuals is decisive yet demanding (Straus & McGrath, 1994). According to studies by McGrath (1993), for task and media fit on information richness, the face-to-face channel is ideal for negotiating conflicts of interest. Negotiations require media richness for task success, which face-to-face offers as a medium (Suh, 1999). This is followed by video with a marginal fit as information transfer tends to be too lean, and audio and texting with a reduced compatibility as these channels are more reserved.

2.2.2 Framework of Integrated Marketing & Grounding

The Theoretical Framework of Integrated Marketing Communication by Kitchen & Schultz (1999) transferred to the context of negotiations blends channel components to equalize deficiencies of one type of media with the performance of another and this channel symbiosis constructs a cohesive structure. The concept of Integrated Marketing Communication searches applicable and effectual media amalgamations (Bothorel, 2021) to strengthen rapport among individuals. The theory of symbiosis in marketing communications is an impactful harmonization of communicative content. This effect draws from the recipient receiving a consistent message from various media (Wang, 2006). Implementing consistency in content transmission requires organizations to comprehend recipients' perception of media (Danaher & Rossiter, 2011). Research on the symbiosis of conventional media (Edell & Keller, 1989; Naik & Raman, 2003) and symbiosis of classic and digital media (Chang & Thorson, 2004; Dijkstra et al., 2005) support the concept of synergy in multi-channel negotiations. Based on Clark and Brennan (1991), different features are present in media channels (face-to-face, video, phone, and email), establishing grounding in communication. Grounding in communication is the synergistic progression to establish a common understanding among different parties (Baker et al., 1999; Clark & Wilkes-Gibbs, 1986). According to Clark and Brennan (1991), the following media characteristics - condensed for this study - can develop using various channels: a mutual presence, vision and acoustics, transitions, synchronism, successions, and revision.

2.2.3 Uses and Gratification & Media Synchronicity Theory

On the one hand, the Uses and Gratification Theory implies that among individuals' demands for "information, knowledge, and understanding" exist (Danaher & Rossiter, 2006, p. 5; Katz et al., 1973; Nysveen et al., 2005). Following this concept, a negotiation comprises different requirements such as information, knowledge, and understanding. This aligns with contemporary discoveries of Voeth et al. (2022) that information and moderate targets (Backhaus & Voeth, 2010) are the key for negotiation success. On this basis, use of various media seems obligatory in a negotiation. Different communication channels can meet these needs to varying extents. On the other hand, media synchronicity deals with media choice in the sense of choosing media for performance (Dennis & Robert, 2009). According to the Media Synchronicity Theory, communication aims at creation of a mutual comprehension between individuals (Miranda & Saunders, 2003; Rogers, 1986; Te'eni, 2001). The communication participants can co-construct a shared understanding (Miranda & Sanders, 2003; Weick, 1979); however, arriving at a mutual understanding cannot be taken for granted and requires effort (Sitkin et al., 1992). Agreeing on meaning is achievable yet not a prerequisite to an interaction. A transmission of communicative content can modify comprehension, yet it is not a compulsory achievable. In essence, an interaction entails "a process in which participants create and share information to reach a mutual understanding" (Jiang, 2019; Rogers, 1986, p. 199).

Distributing information equates in an interchange of cognitive encoding and emerging understanding. A cognitive encoding entails not only comprehension yet also deciphering content. On the one hand, conveyance procedures transmit a diversity of novel information for formation and review of a cognitive model. On the other hand, convergence procedures are the elaboration on formerly treated content concerning a prior assessment of a setting. The goal is to harmonize views regarding informative content, which requires consensus among parties to establish mutual agreement or to confirm an impossible consent (Lind & Zmud, 1991). Like common processes, negotiation involves, in different quantities and intervals, conveyance and convergence progressions depending on the context (Dennis et al., 2008). Nevertheless, these two processes have fundamentally different implications for media choice.

2.3 Social-Psychological Perspectives

Strategic communication theories analyze media characteristics and tasks, whereas social-psychological perspectives focus on the communicator–medium interaction (Geiger, 2020). A communication medium can aid or distract social processes essential to communicators. In due course, social-psychological perspectives complement the theoretical grounding work for multi-channel negotiations. According to Geiger (2020), relevant theories for assessment in the context of multi-channel negotiations are the Social Presence Theory (Short, 1974), the Barrier Effect by Lewis and Fry (1977) and Carnevale et al. (1981), the Social Information Processing Theory (Walther, 1992, 1994, 2015) and the Communication Orientation Model (Swaab et al., 2012) with confirmation of the concept for bargaining by Bollen & Euwema (2013), Geiger (2020), Haselhuhn et al. (2014), and Nijstad & De Dreu (2012). Geiger (2020) outlines that the Social Presence Theory and Social Information Processing Theory are utilized broadly in various communication settings, while the Barrier Effect and Communication Orientation Model solely apply to negotiations. Social Information Processing Theory and the Communication Orientation Model include boundary conditions and the negotiator attitude (Barkhi et al., 1999, 2004; Geiger & Parlamis, 2014; Geiger, 2020; Huang et al., 2008; Wang & Doong, 2014) beneficial to multi-channel negotiation research. The following sections outline necessary content of the described theories, which are of relevance to this research.

2.3.1 Comparing Social Presence with Psychological Distance

The concept of social presence (Short et al., 1976) and psychological distance (Citera et al., 2005; Stuhlmacher & Citera, 2005) both attribute the capacity to create social presence only to the medium of face-to-face.

Table 3. Social Psychological Perspectives

Social Psychological Perspectives	Author
Barrier Effect	Lewis & Fry (1977) Carnevale et al. (1981)
Negotiator Attitude	Geiger & Parlamis (2014)
Psychological Distance Perspective	Citera et al. (2005) Stuhlmacher & Citera (2005) Dorado et al. (2002) Kiesler et al. (1984) Morris et al. (2002)
Reduced Social Cues	Kiesler & Sproull (1992) Sproull & Kiesler (1986)
Social Information Processing Theory	Walther (1992, 1994, 2015)
Social Impact Theory	Latané (1981)
Social Presence Theory	Short et al. (1976) Spears & Lea (1994)

Note: This compilation of social psychological perspectives does not present all applicable and available theory. It is only a selection of relevant theories in relation to the topic of multi-channel negotiations.

Related theoretical reasoning labels this phenomenon a state of reduced visual indicators (Kiesler & Sproull, 1992; Sproull & Kiesler, 1986), described and reviewed within the scope of social impact (Latané, 1981). Diminished individual presence and expansion of distance is the product of absence of manifold channel attributes (Dorado et al., 2002; Kiesler et al., 1984; Morris et al., 2002; Spears & Lea, 1994; Sproull & Kiesler, 1986; Stuhlmacher & Citera, 2005). This state can lead to less cooperation in negotiations due to psychological distance and frail rapport.

Social presence (Short et al., 1976) entails individual prominence in an exchange. The unavailability of cues ultimately leads to reduced prominence of the interacting person. The counterpart is less attentive to the participant's presence (Williams, 1977). In addition, Kiesler and associates (1984) and Siegel et al. (1986) describe the prevalence of unrestrained interactions in digital settings as opposed to an exchange with visual access. Virtual space even encourages offensive interactions (Baltes et al., 2002; Stuhlmacher & Citera, 2005). Overall research by Stuhlmacher & Citera (2005) suggests enhancements in consensus building paired with high satisfaction while using visual access such as face-to-face as opposed to digital settings. Theories on psychological distance imply augmented depersonalization in the digital space (Daft & Lengel, 1984; Latane, 1984; Trevino et al., 1990; Wellens, 1986, 1989; Williams, 1977) as channel capacity to convey content is unequal (Wellens, 1986, 1989). According to this understanding, intimacy is established with face-to-face interactions, whereas the digital sphere fosters personal detachment (Stuhlmacher & Citera, 2005). The next section discusses the theory of the Barrier Effect and its relevance to multi-channel negotiations.

2.3.2 The Barrier Effect

As most negotiators render a negotiation competitive (Neale & Bazerman, 1992; Pruitt & Carnevale, 1993), the Barrier Effect is a concept that tackles this challenge. Theory on psychological barriers by Lewis and Fry (1977) and Carnevale et al. (1981) imply a barrier to avoiding visual dominance and improving negotiators' cooperativeness (Pruitt, 1981). Reduced adverse conduct and cooperative outcomes are documented between negotiators when removing optical entry instead of seeing the opponent visually (Carnevale et al., 1981; Carnevale & Isen, 1986; Lewis & Fry, 1977). However, the authors emphasize that the effects of the barrier are not caused by the overall context of competition (Pruitt, 1981). Conversely, in integrative negotiations, the barrier does not act favorably (Stuhlmacher & Citera, 2005). To minimize hostile behavior, audio or written exchange is advised as opposed to video conferencing and face-to-face, which increase competitive behavior (Dalkey et al., 1969; Martin & Nagao, 1989; Delbecq & Van de Ven, 1971).

It can be problematic to derive a recipe for successful multi-channel negotiations from strategic communication theories as these are not only focused on negotiation. The theoretical perspectives discussed herein concerning multi-channel negotiations require insights into media combinations in parallel or sequence

(Damian, 2002). Media Richness and Grounding Theory exclude boundary conditions and ensembles of media. On the contrary, Media Synchronicity Theory introduces boundary settings such as task or partner familiarity and recommends a mix of media for communication subprocesses leading to optimal conveyance and convergence of information. As the strategic character of this research is the impact assessment on effectiveness and efficiency by negotiation behavior in multi-channel negotiations, a concept which encompasses this relationship is nonexistent. A model is needed covering all these topics relevant to this study. The preceding chapter theoretically discusses behavioral impacts on the multi-channel negotiation progression and outcome.

2.4 Behavior – The Driver Behind Negotiation Performance

Since the beginning of research on bargaining, the quest for formulas and control levers, which explain the success or failure of negotiations, has been of utmost importance. However, "views of rational media choice and task performance do not explain performance" (Valacich et al., 1994, p. 18). In line with Walther (1992), rather than focusing on media characteristics, a closer examination of negotiators using media promises the needful insights. Over the long run, communication transcends media (Walther, 1992), which contradicts McLuhan's theory that the channel holds the content of the transaction (McLuhan, 1964). To such a degree, success in a negotiation is attributed, among other aspects, to the heterogeneous behavior of negotiation parties (Albers & Hildebrandt, 2006). Subsequently, this research validates how success in a multi-channel mix depends on the behavior of negotiators. Consequentially, it is imperative to analyze negotiators using media and examine behavior and communication in the negotiation. Negotiators decide when and how information is exchanged effectively and efficiently through which negotiation channels, and with which negotiation partners (Ambrose et al., 2008; Arthi, 2009; Dennis et al., 2008; Dennis & Robert, 2009; Lasswell, 1948; McLuhan, 1964; Olekalns et al., 2003; Swaab et al., 2012; Thompson, 2009; Walther, 1992, 1994).

2.4.1 Measurement of Negotiation Behavior

There is a necessity to focus on empirical measurement of behavior. This scientific approach of identifying interrelations through quantifications stipulates understanding regarding drivers for various forms of bargaining interactions. Measures of behavior and performance help classify a particular behavior for successful negotiations (Stoshikj, 2014). Thompson (1990) defines two measures for negotiation behavior: economic and social-psychological. The economic measure is outcome-oriented due to a specific behavior (Nash, 1950). The social-psychological measure combines the resulting outcome with properties of the bargaining phase and relates to social perception of the negotiation. The negotiator cannot always objectively analyze the negotiation situation. Conforming to this state, the negotiator makes assumptions and conclusions depending on the situation and perceived behavior. In line with Thompson (1990), respondents of this survey assess their social perceptions of their behavior during the negotiation. In conclusion, the economic and social-psychological measures influence and complement each other. However, negotiation research uses economic measures more widely for behavioral assessments than social-psychological ones. Nevertheless, the latter integrates into this analysis and research, as in many situations, negotiators only have their perceptions to draw conclusions from, and it is helpful to equate those subjective discernments with unbiased quantification. This study unites economics with social-psychological measurement. The economic measurement of this study analyzes negotiation effectiveness and efficiency. Assessment of bargaining success grounds in social-psychological quantification of rating bargaining behavior during the negotiation process (Mannix et al., 1995; Weingart et al., 1993).

2.4.2 Behavior in Multi-Channel Negotiations

Increasingly, negotiators favor manifold media, employ these, and even shift in media application (Neslin & Shankar, 2009). Behavioral effects occur – so-called reactions – in individuals, dyads, or groups because of using multiple media (Lowery & DeFleur, 1988; Perse, 2001; Tsfati, 2011). A multiple media usage impacts behavior and creates behavioral change (Wangenheim & Bayon, 2014). In the context of negotiations, these findings indicate negotiation behavior becoming more complicated with augmented media use. Existing research frequently centers on negotiation behavior in solitary channels and isolation from alternative media (Yu et al., 2011). Moreover, deviations in findings occur in negotiation behavior concerning a specific negotiation channel. Therefore, behavior of negotiation parties is analyzed while using multiple communication channels. Latent constructs reveal the bargaining effects on the success of the interaction. The measurement of behavior and performance helps to identify a specific behavior as being responsible for an effective and efficient negotiation outcome. This approach assumes that negotiation channels are the determinants of negotiator behavior in this study. Opposed to this assumption, several authors report no differences in behavior in general during the use of various communication media (Delaney et al., 1997; Staples & Zhao, 2006; Zornoza et al., 2002). For negotiations, no disparity in behavior using various media channels has been identified (Delaney et al., 1997; Geiger, 2020; Griffith & Northcraft, 1994; Yang, 2012; Zornoza et al., 2002). Opposed to these prevalent scientific findings, this study hypothesizes that multi-channels impact negotiation behavior. This research challenges prevalent scientific findings in a variable interaction analysis and validates corresponding outcomes with existing research.

2.4.3 Measurement of Multi-Channel Behavior

Grounded in findings of the systematic literature review, an abstract construct (Rossiter, 2008) is needed for this study on multi-channel negotiations to discuss higher-level behavioral negotiation strategies. Specifically, a latent construct can contribute to analyzing unobservable behavioral drivers (Coltman & Devinney, 2007) for negotiation success. Understanding the interrelation between the latent behavioral constructs and negotiation success is required. Moreover, the conceptual model for this research study must fulfill the requirements of real-life complexity in multi-channel negotiations (Kersten & Lai, 2007; Schoop, 2010; Watkins, 1999). To this extent, behavioral dimensions in multi-channel negotiations are explored with this conceptual framework to understand their impact on negotiation performance. This chapter outlines essential terminology and constructs used in this study. After the definition of the terminology and constructs, it is possible to assess how the concept can be influenced positively and if a positive relationship exists between the variables, especially concerning the success of a negotiation.

Additionally, this study uses data from negotiation experts in companies who evaluate behavioral success factors in multi-channel negotiations. These behavioral success factors are specified in the questionnaire via measurement models with multiple items (Churchill, 1979; Bergkvist & Rossiter, 2007). This approach increases the ambition to search for specific behavioral patterns in a negotiation (Thompson, 1990). Such a scientific method aims to mature into a clarifying structure that assists practitioners and academics to comprehend heterogenous interrelations in behavior during bargaining phases. An exhibit of disparities in interactions tackles a significant challenge in negotiation research: creating a framework for exploring modes of bargaining behavior, which attain acceptable outcomes to comprehensively understand the interrelations in behavior.

By dividing the construct of behavior in multi-channel negotiations into dimensions, the content of this construct is theoretically deducted (Albers & Eggert,

1988). After the definition and understanding of the causal relationships in the construct, an assumption is made on how the construct is influenced positively or negatively regarding its outcome. Under these circumstances, behavioral effects in multi-channel negotiations are assessed, influencing negotiation effectiveness and efficiency through four behavioral dimensions and three moderator variables from the theoretical assessment. Behavioral dimensions in the scope of this study are self-efficacy, agility, synchronicity, and behavioral control. Corresponding moderating effects are complexity, cognitive resource demands, and performance risk.

Self-Efficacy

Self-Efficacy originates in the Reasoned Action Theory (Bandura, 1977). It is the belief one can accomplish goals or achieve performance standards (Judge et al., 2002). This definition aligns with the Theory of Expectancy by Vroom (1964), which deals with one's belief that one's effort results in projected outcomes (Volkema et al., 2013). Task familiarity is a driver behind manifesting success (Bandura, 1997, 2001). Intense manifestations result in encouragement to struggle (Volkema et al., 2013). In this sense, for a modification of behavior self-efficacy is required as it reflects people's confidence and motivation to execute perseverance with performance (Bandura, 1977, 1982; Kirk & Brown, 2003; Lim & Weissmann, 2021). Kirk and Brown (2003, p. 46) identified that self-efficacy "explained the most variance in performance" (Volkema et al., 2013). Accordingly, self-efficacy impacts the process and outcome (Brett et al., 1996; Gist et al., 1991; O'Connor & Arnold, 2001; Stevens & Gist, 1997; Sullivan et al., 2006; Volkema et al., 2013). In this research, self-efficacy is defined as the bargainer's belief in owning the capability to evaluate negotiation channels, select these independently, and implement these effectively and efficiently. It depends on the individual judgment regarding one's capabilities to select suitable negotiation channels and not on the ability to judge one's negotiation skills. Instead, individuals would only initiate interactions they are able to handle (Volkema et al., 2013). In the next section, the behavioral dimension of agility is presented and defined for the context of this study.

Agility

Agility in this study is a fused structure of flexibility, which deals with heterogenous and ambiguous settings (Charbonnier-Voirin, 2011). Various terms label these capabilities as characteristics of agility (Dove, 2001; Herrmann et al., 2006; Kidd, 1994; Sharifi et al., 2001; Sharifi & Zhang, 1999). To narrow down essential and interrelated capabilities within the scope of this study, it is necessary to only consider individual agility as opposed to organizational agility. Individual agility is flexibility and adaptive thinking and behavior, especially in response to changing circumstances (De Meuse et al., 2010). It is the aptitude to promptly attain new skills, adapt to new situations, and respond to challenges with resilience (Park & Park, 2021). In this context, individual agility reflects the aptitude to react to uncertainty and transformation by processing information (Marschak & Reichelstein, 1998; Muduli, 2017; Nafei, 2016a, 2016b; Pitafi et al. 2018; Qin & Nembhard, 2015; Tushman & Nadler, 1978). For this, agile individuals acquire various information from multiple places (Alavi et al., 2014; Cai et al., 2018; Daft & Lewin, 1993; Sheppard & Young, 2006). Hereby, in-time reaction transforms challenges into potential prospects (Alavi et al., 2014; Pitafi et al. 2018; Sherehiy, 2008). For that reason, agility in the context of negotiation, is reactive and rapid bargaining in accordance with negotiation targets and the capacity to align resources to an evolving setting. Instinctively, agile negotiators react quickly and effectively to changing negotiations resulting in customized negotiations. Agility refers to a negotiator's ability to accelerate activities on critical paths. It indicates a time-based task effectiveness (Arteta & Giachetti, 2004; Kumar & Motwani, 1995; Voeth et al., 2020). Studies investigating development of such agility in negotiations are infrequent (Cai et al., 2018; Hosein & Yousefi, 2012; Sherehiy & Karwowski, 2014). This study synthesizes characteristics of the agile negotiator. Within the scope of this study, the definition of agility in multi-channel negotiations is that the negotiator negotiates agile using diverse communication channels during the negotiation, which allows fast and effective reactions towards unexpected changes. Multi-channel agility, as a processual negotiation behavior, is identified through the four components: reaction (Lin et al., 2006a, 2006b; Sharifi & Zhang, 1999), competence (Dove, 2001; Shafer, 1997), flexibility (Sherehiy et al., 2007) and speed (Dove, 2001; Raiffa, 1982; Shafer, 1997). The leverage effect of multiple channels during the management of negotiation interactions produces negotiation agility. Through interpretation and application of in-time access to relevant information (Alavi et al., 2014; Daft & Lewin, 1993; Hopp & Oyen, 2004; Marschak & Reichelstein, 1998; Nafei, 2016a, 2016b; Sherehiy, 2008; Sumukadas & Sawhney, 2004; Tushman & Nadler, 1978) agile negotiation occurs via multiple negotiation channels. This information base enables a rapid and flexible reaction to changes (Brown & Agnew, 1982) by focusing on negotiation goals and the competence of utilizing resources efficiently. Inevitably, agility is an effect of the use of multi-channel negotiations, which impacts negotiation effectiveness and efficiency. It is assumed that negotiator agility can be improved through mental conditioning.

Studies reveal that exposure to the dynamics of social media support in developing individual agility (Cai et al., 2018). Depending on the individual and context, alliances as well as independence can act favorably on agility (Sherehiy & Karwowski, 2014). These alliances enable access to concise information in the right time resulting in agile behavior (Alavi et al., 2014; Cai et al., 2018; Eshlaghy et al., 2010; Pitafi et al., 2018; Vinodh & Prasanna, 2011). Factors for fine tuning agility are the comprehension of emotions and social context (Hosein & Yousefi, 2012). Research findings strongly support the improvement of agility through the above factors (Alavi et al., 2014). In the next section, the behavioral dimension synchronicity is introduced and defined in the context of this study.

Synchronicity

Media Synchronicity Theory matches the communicator's synchronicity with media synchronicity through “conveyance and convergence processes” (Dennis et al., 2008; Parlamis & Geiger, 2014, p. 362). Information conveyance and meaning convergence are required to achieve collective understanding (Parlamis & Dibble, 2019). Information is transported to the receiving end through a conveyance process to construct a cognitive understanding of the content (Geiger, 2021; Schmidtke & Cummings, 2017). This information is discussed and interpreted in a convergence process (Espinosa et al., 2015). Communication, especially negotiation, consists of subprocesses in which the amount of each conveyance and convergence differs (Geiger & Laubert, 2018). Each subprocess in a negotiation requires a different channel for successful conveyance and convergence. Synchronicity is a coordinated, synchronous behavior with a shared goal to create a win-win situation or to end in an integrative negotiation (Geiger, 2021). It is the ability to work simultaneously with coordinated behavior (Dennis et al., 2008). A construct of synchronicity analyses whether the content and meaning of a multi-channel interaction can create a mutual understanding to achieve a sustainable result (Dennis et al., 2008; Park et al., 2014; Ryoo & Koo, 2010). Due to its release date, the concept of Media Synchronicity Theory is less tested by the academic community than Media Richness Theory or Grounding Theory. This validation gap offers an imperative opportunity to test the impact of synchronicity in multi-channel negotiations within this study. For this study, synchronicity in multi-channel negotiations is defined as a coordinated, synchronous behavior of the negotiation parties – caused by using diverse negotiation channels – with a shared goal to create a win-win situation, which can lead to an integrative negotiation. The construct synchronicity contemplates if a multi-channel interaction's information or semantic content in a negotiation process produces a mutual negotiation conception to obtain a sustainable negotiation deliverable. The efficacious matching of negotiation channels with the negotiation content determines if synchronicity can be established, positively impacting negotiation effectiveness and efficiency. In the next section, the behavioral dimension of behavioral control is presented and defined in the context of this study.

Behavioral Control

From Self-Efficacy Theory, Bandura (1977) developed the Theory of Behavioral Control in communication (Lim & Weissmann, 2021). It was Ajzen (1991) who introduced the Theory of Reasoned Action to discuss findings on observed behavioral control (Ajzen, 1991, 2012; Fishbein & Ajzen, 2010; Trenz, 2015) and upgraded it to the Planned Behavior Theory (Lim & Weissmann, 2021). An assessment of control on behavior is the extent to which one can regulate performance (Bandura, 1977; Lim & Weissmann, 2021; Terry & O'Leary, 1995). A behavioral control mechanism depends on information, experiences, projected support, and obstacles (Lim, 2003). Behavior amplifies with supplementary assets, prospects, fewer hindrances, and hurdles. Control is often intrinsic to the individual through obtained abilities, capacities, determination, and needs. It can also be formed by peripheral factors such as time, opportunities, or reliance on others (Ajzen, 1985). In essence, perceived behavioral control directly forms behavior (Armitage & Conner, 2001; Cooke & French, 2008; Yzer, 2012). Understanding a realistic level of possible regulation defines the level of control on behavior (Ajzen, 2002; Lim & Weissmann, 2021), which expedites or inhibits conduct. Transferring the Theory of Behavioral Control (Collier & Sherrell, 2010) to negotiations equates to perceived control over objects or processes. It reflects the negotiator's confidence in correctly assessing the required performance behavior. Within this study, regulating behavior outlines as the negotiator's control over objects and processes while using multiple channels. It is assumed that negotiators prefer media channels over which they believe they have more control instead of channels where perceived control is less to positively impact negotiation effectiveness and efficiency. The next section deals with the selection of dependent outcome and moderating variables for the conceptual model of this research.

2.5 Negotiation Effectiveness & Efficiency

A negotiation outcome is appraised as centered on economic criteria divided into effectiveness and efficiency (Heinen, 1976; Macharzina & Wolf, 2008;

Schmidt, 2020). Non-economic criteria are not evaluated as the focus of this study is to enhance economic outcomes in managerial negotiations (Hyder & Prietula, 2000). The endogenous variables of effectiveness and efficiency are frequently used in interdisciplinary and negotiation research (Katz & Kahn, 1978a, 1978b, 1978c; Neely et al., 2005; Purdy et al., 2000). On the contrary, dependent variables analyzed in reviewed studies (Geiger, 2020) show increased research interest in socio-emotional outcome variables over decades. The most prominent dependent variable category, economic profit, is replaced by negotiation behavior from the 90s. From 2010, the three most relevant dependent variables in academic research are 1) joint profit, 2) negotiation behavior, and 3) satisfaction (Carnevale et al., 1981; Carnevale & Isen, 1986; Geiger, 2020; Stuhlmacher & Citera, 2005). Nonetheless, this study analyzes negotiation behavior as an independent variable with its effect on the performance measures of negotiation effectiveness and efficiency. To this degree, negotiation effectiveness stands for the ability to select adequate negotiation targets (Backhaus & Voeth, 2010) and to negotiate these in quantity and quality competently and effectively to achieve a desired negotiation outcome sustainably (Barisch, 2011; Bazerman et al., 1992; Butler, 1999; Drucker, 1974; Groves et al., 2015; Movius, 2008; Purdy et al., 2000). The causal relationship manifests itself through the selection of communication channels which cause effects on the behavioral dimensions of self-efficacy, agility, synchronicity, and behavioral control, thus impacting negotiation effectiveness. The measurement scale for negotiation effectiveness is developed by Barisch (2011), focusing on target achievement (Butler, 1999; Cunningham, 1977). This focus on target achievement prioritizes actions to do the right things in a negotiation leading to effectiveness (Drucker, 1974; Purdy et al., 2000). As a result, both short- and long-term impacts of effectiveness are assessed (Lax & Sebenius, 1986). A multi-channel negotiation is efficient if it is conducted goal-oriented, structured, and cost-efficient to attain the postulated negotiation targets sustainably. It is assessed in this construct if resources (e.g., monetary investment and time) are used efficiently (Butler, 1999; Voeth et al., 2006) in the multi-channel negotiation to reach a negotiation target. Through a selection of media, the behavioral effects of self-efficacy, agility, synchronicity, and behavioral control emerge, which impact negotiation efficiency. As a result, the conduct of doing things correctly is the fundamental understanding of efficiency (Drucker, 1974).

2.6 Moderator Variables

It is imperative to consider moderating effects in complex causal structures (Eggert et al., 2005) to assess under which circumstances a causal relationship is especially strong or distinctly weak. Within the scope of this research, it is assumed that the causal relationship is fragile under certain circumstances. Moderator variables, which impact the main effects negatively and weaken these, are complexity, cognitive resource demands, and performance risk. The above moderating effects are not part of the causal relationship yet influence the impact path and strength between independent and dependent variable relationships (Baron & Kenny, 1986; Sheer & Rice, 2017). Anticipated correlations among variables modify if the moderator variable changes. Firstly, the moderation of complexity on multi-channel behavior is discussed, secondly, the moderating variable cognitive resource demands is presented, and thirdly, performance risk is introduced with all three moderators labeled as interactive terms. Interaction terms are applied (Hayduk, 1987) to test structural equations for a presence of moderation, as all relevant moderators in this study have a continuous scale (Jaccard & Wan, 1996).

2.6.1 Complexity

Conceptualizing and operationalizing the complexity construct includes a multidimensional approach, which needs a specification (Braun & Hadwich, 2016). Within the marketing and management context, a multitude and diversity of characteristics of complexity are detected, which constitute facets of this construct. Complexity in multi-channel negotiations references characteristics in the perception of the negotiation expert, leading to high complexity. It relates to how complicated a specific task is believed to be. For example, the positive impact of a behavioral dimension is weakened due to the complexity of negotiation objects. Substantially, as one facet of the complexity construct, this study assesses if the negotiation objects are considered complex. In multi-channel negotiations, according to Braun & Hadwich (2016) it is defined as a multitude of channels with numerous and diverse relationships between the channels. The complexity facet multitude is the perceived quantity of interconnected elements – channels, interfaces, information, and documents – necessary to contribute to negotiation performance and which need to be structured. Variety is a complexity facet that refers to the perceived heterogeneity (Child, 1972) of negotiation channels, the variation in channel properties, the diversity of interfaces between elements and phases, as well as different intensities of the relationship (Morel & Ramanujam,

1999). Interdependence is the perceived interrelation (Bliss, 2000; Blockus, 2010; Cilliers, 2000a, 2000b, 2000c) of individual negotiation channels and the reciprocal relationship between negotiators and interfaces. The complexity dimension dynamic is the perceived change of interfaces, relationships, and dependencies (Senge et al., 1994; Wood, 1986) within the negotiation framework. In the following section, the concept of the moderation effect of cognitive resource demands on the interdependence between behavior and negotiation results is explained.

2.6.2 Cognitive Resource Demands

In this study's context, the "conduit metaphor" (Putnam et al., 1996, p. 340) is used and treats the medium as an outlet among participants. The conduit metaphor is a concept in linguistics and communication theory that describes how language is often conceptualized as a conduit for transmitting information from one person to another (Reddy, 1979). The channel transfers and processes content, which results in an implication (Miranda & Saunders, 2003; Sitkin et al., 1992). Instead of using the reference to language, the reference to negotiation media as a conduit is applied. With this understanding, an interaction is a one-directed procedure to share content via media with an audience. The target group then processes the information from the media. The conduit metaphor suggests that media functions like a pipe that carries information from one place to another without changing its meaning or being affected by the environment in which it is used. The opponent then unlocks content cognitively to strategically plan a counterstrike (Lee, 1994; Petty & Cacioppo, 1986). This concept mirrors the assumption that cognitive resource demands are essential in encoding a negotiation message. As depicted by Herbst et al. (2017), the negotiator's cognition processes a variety of data on opponents and objects resulting in overloaded cognitive processing abilities (Geiger & Hüffmeier, 2020). Respectively, this study defines cognitive resource demands, derived from Wilcox et al. (2011), as the level of thinking and attention a negotiator devotes to a multichannel negotiation. In the following section, the moderator variable performance risk is elaborated.

2.6.3 Performance Risk

Performance risk for the context of this study derives from the structural contingency perspective (Fernandes et al., 2018; Fry, 1982; Perrow, 1967; Pugh et al., 1969; Thompson, 1967; Van de Ven et al., 1976; Van de Ven, 1985; Woodward, 1965). Transferring this performance risk concept of numerous authors

(Ahuja & Carley, 1998; Daft & Macintosh, 1981; Edström & Galbraith, 1977; Galbraith, 1973; Tushman & Nadler, 1978) to negotiations stands for the ability to perform in an environment of uncertainty. Performance risk is a component of observed risk (Mitchell & Greatorex, 1993; Stone & Gronhaug, 1993), which is the extent to which an object or process performs differently in expected ways (Lutz & Reilly, 1974). Specifically, in the context of multi-channel negotiations, this pertains to the ability to perform as expected and thus satisfy the negotiator's needs. It is the fear of malfunctioning in contrast to a loss in gain (Mitchell & Greatorex, 1993). Within this study, the definition of the moderating variable performance risk is the negotiator's belief that there is a potential level of risk that the multi-channel negotiation may not yield the expected results as grounded in work of Keh & Pang (2010) and Stone & Grønhaug (1993).

2.7 Hypotheses and Conceptual Model

Statistical analysis tests the postulated hypotheses of this study for plausibility, and knowledge outside of statistics is also required (Backhaus et al., 2015). According to the Hempel Oppenheim Schema (Hempel & Oppenheim, 1948), the deductive nomological approach tests the hypotheses below, which are grounded on theoretical, factual, and logical considerations. According to findings in scientific literature, it is assumed that behavioral dimensions can positively impact the negotiation outcome. Beyond doubt, all behavioral dimensions could negatively impact negotiation effectiveness and efficiency. For reasons of simplicity, no additional hypothesis covers this scenario. Centered on Bandura (1997) and Pekkala & Van Zoonen (2021), the applied self-efficacy theory in this research is the confidence in the ability to evaluate negotiation channels, select these independently, and implement these effectively and efficiently. In accordance, it is assumed that self-efficacy positively impacts negotiation effectiveness and efficiency.

H1a Self-efficacy positively impacts negotiation effectiveness.
H1b Self-efficacy positively impacts negotiation efficiency.

Agility in a negotiation is "the capacity to react quickly to rapidly changing circumstances" (Brown & Agnew, 1982, p. 29), which demands concentration on negotiation objectives and matching resources to the demands of changing negotiation circumstances. As follows, agility refers to a negotiator's ability to accelerate activities on critical paths. It is a direct indicator of a negotiator's time-based effectiveness (Kumar & Motwani, 1995) in the negotiation. In accordance, hypothesizing that agility results from multi-channel negotiations, which impacts the negotiation effectiveness and efficiency, it is argued that:

H2a Agility has a positive effect on negotiation effectiveness.
H2b Agility has a positive effect on negotiation efficiency.

Synchronicity - caused by using diverse negotiation channels - is a coordinated, synchronous behavior of the negotiation parties with a shared goal to create a win-win situation, which can lead to an integrative negotiation (Arthi, 2009; Dennis et al., 2008; Dennis & Robert, 2009). The construct synchronicity contemplates if a multi-channel interaction's information or semantic content in a negotiation process produces a mutual negotiation conception to obtain a sustainable negotiation deliverable. The efficacious matching of negotiation channels with the negotiation content determines if synchronicity is established in the negotiation, which positively impacts negotiation effectiveness and efficiency. Respectively, it is hypothesized:

H3a Synchronicity has a positive effect on negotiation effectiveness.
H3b Synchronicity has a positive effect on negotiation efficiency.

Within this study, behavioral control is defined as feeling control over objects and processes (Lim, 2003) of the negotiation while using multiple channels. Negotiators prefer media channels over which they believe they have more control instead of channels where perceived control is less to impact negotiation effectiveness and efficiency positively. Theorizing on this interrelation, the subsequent hypothesis is derived:

H4a Behavioral control positively impacts negotiation effectiveness.
H4b Behavioral control positively impacts negotiation efficiency.

It is assumed postulated relationships of behavioral dimensions and the negotiation outcome depend on the values of third variables (Ramayah et al., 2018). A priori, it is hypothesized that moderating variables negatively affect the direction and intensity of all construct relationships in this model (Dawson, 2014).

H5 – H7 The effect of behavioral dimensions on negotiation effectiveness and efficiency is decreased through moderator variables complexity, cognitive resource demands and performance risk.

The structure and content of the negotiation process displayed in this contextual framework are derived from Ghauri & Usunier (2003), Thompson (1990), and

Weiss (1997). The context for negotiation behavior stems from Purdy et al. (2000), Sheffield (1995), and Swaab et al. (2012), who declare communication media as the negotiation context for behavior. Constructed around research of Yu et al. (2011), a heterogenous negotiation behavior develops when negotiators syndicate manifold channels. Further, the arrangement and substance of the negotiation result follow the approach of Barisch (2011), Bercovitch & Jackson (2001a, 2001b), Pinfari (2012), and Sawyer & Guetzkow (1965). Incontrovertibly, the result of heterogeneous behavior affects effectiveness (Pookulangara et al., 2011; Pookulangara & Natesan, 2010; Wang et al., 2015; Yu et al., 2011) and efficiency (Neslin & Shankar, 2009). To test for possible moderating effects, the variables complexity, performance risk, and cognitive resource demands are included into the model. Ideally, a multi-channel strategy can be derived when we match the empirical study results with the conceptual framework's structure. The subsequent section presents the applied methodology of this study.

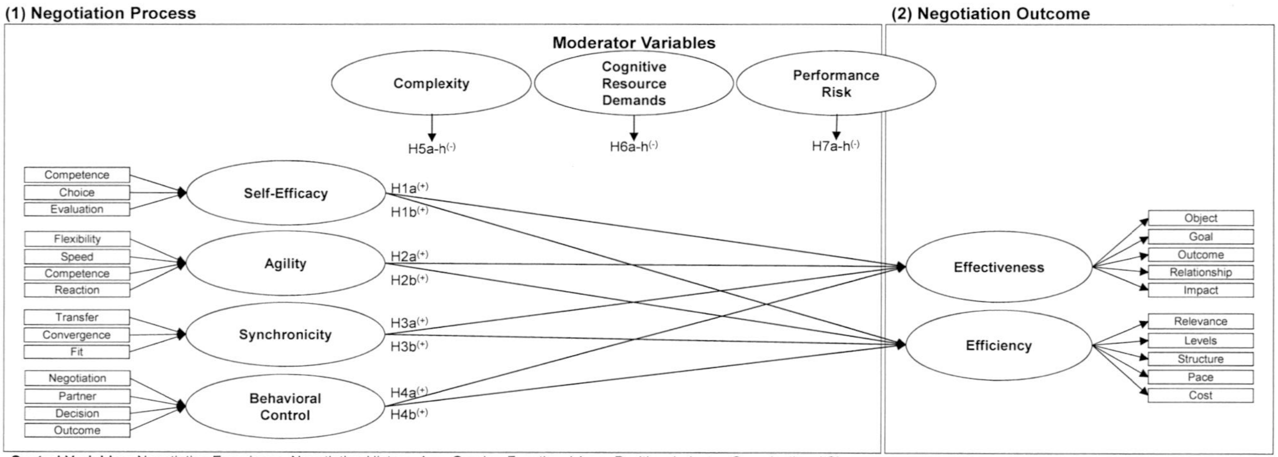

Control Variables: Negotiation Experience, Negotiation History, Age, Gender, Functional Area, Position, Industry, Organizational Size

Figure 4. Conceptual Framework on Behavior in Channel Negotiations

3 Methodology

In a two-step approach, existing research on media use in negotiations as secondary data was analyzed in a systematic literature review, which developed in medicine and healthcare in the 1970s. This concept developed by Chalmers and colleagues (1993) of systematically identifying, evaluating, and synthesizing all available research evidence relevant to multi-channel negotiations was applied to provide an unbiased and comprehensive summary of the present understanding on this topic. Findings of the qualitative literature review were processed to examine the impact of multi-channel in business negotiations, and from this, a concept for quantitative measurement was extracted. A quantitative data analysis was applied to validate qualitative research findings in negotiations. The analysis of negotiation behavior in multi-channel negotiations was established on negotiation experts' experiences and assessments collected through an online survey to generate primary data.

3.1 Systematic Literature Review & Quantitative Analysis

In a first step, an extensive literature review analyses the emergence and evolution of multi-channel negotiations covering 30 years of academic publications. Search terms utilized are communication AND negotiation or e-negotiations, to name a few. Articles and doctoral dissertations were retrieved from the catalogues of Business Source Premier (through EBSCO), Google Scholar, Jstor, ScienceDirect, SAGE journals, and SpringerLink. As a limitation, only article findings concerning negotiation are processed. This excludes arbitration, conflict resolution, consensus building, or symbiotic bargaining games. Meta-analyses with a focus on bargaining channels (Baltes et al., 2002; Stuhlmacher et al., 2007; Stuhlmacher & Citera, 2005; Swaab et al., 2012) offer additional negotiation media insights for investigation. Theoretical perspectives were included if used in empirical studies. Through the literature analysis negotiation behavior is identified as an independent variable and the negotiation outcome as a dependent variable (Backhaus et al., 2016; Swaab et al., 2012; Yuan et al., 2003). In the second step, research results regarding positive and negative behavior were analyzed, and relevant behavioral dimensions were selected to measure their influence on effectiveness and efficiency in multi-channel negotiations. The impact of behavioral dimensions was measured through self-efficacy, agility, synchronicity, and behavioral control.

Additionally, three moderator effects were examined: complexity, cognitive resource demands, and performance risk. Based on theoretical and logical considerations, hypotheses regarding effects of behavioral dimensions on the negotiation outcome were formulated. Negotiation data was collected from 733 international negotiators between September 2022 to May 2023 to survey multichannel negotiations. The evaluation of interrelationships of behavioral dimensions on the outcome of multi-channel negotiations was conducted in a structural equation model.

3.2 Survey Methodology & Data Collection

For the survey methodology, a format of questioning was selected to obtain information about behavior in multi-channel negotiations following standardized procedures, according to Churchill (1979). The questionnaire was a partially standardized online survey developed with the survey software Unipark, a product of Tivian XI GmbH. The questions employed were closed questions, ranking scales, and open questions. With basic queries as those included in the study survey, where information is distinct, a common-method bias is negligible. Furthermore, questions regarding demographic data were listed at the end of the questionnaire. The processing time for the questionnaire varied between 13-20 minutes for 27 questions. A survey pre-test was conducted with 30 participants as a pre-study to substantiate the questionnaire for comprehensibility (Weisberg, 2008). The data collection process started in September 2022 and was concluded in May 2023. For international acquisition, potential participants were contacted with personalized messages through mailing, networking, personal contacts, or via the professional networking platforms XING & LinkedIn and incentivized with results of the study. The sample involved international cross-industry practitioners operating in various continents globally with negotiation experience and a strategical outlook due to their prominent organizational positions to ensure dependable and authentic outcomes. The sampling comprises 733 global B-to-B negotiation practitioners from diverse industries. It was mandatory that experts had negotiated in the past 12 months to receive access to the survey. An exit filter only considered input from negotiation experts involved in a negotiation in the past 12 months to safeguard the quality criteria of the actuality of statistics. This approach permitted negotiators to only share details connected to their last relevant negotiation. Participants were asked to mentally transfer into their last relevant negotiation to use retrospective data. For a concluding statistical analysis of survey data, a total of 733 questionnaires were used. Survey respondents represent 24% top-level managers (e.g., managing director, board member, executive), 21% middle-level managers (e.g., senior department head, division, or area manager), 11% first-level managers (e.g., team or group leaders), 14% managers without a team, 11% administrators, team members or assistants, and 16% named an alternative position.

3.3 Operational Measures & Causal Analysis

After introducing the experimental design, information regarding the operationalization of constructs, the applied scales, the procedure of data analysis, and quality criteria were introduced. Modeling the measurement took place on a high abstraction level examining hierarchies between relationships of basic behavioral constructs using measurement theory (Churchill, 1979). According to Homburg & Giering (1996), construct development occurs through careful content-related considerations instead of pure data analysis. At first, for the development of a measurement model in the conceptualizing phase, a definition of the constructs was developed (Braun & Hadwich, 2016). Secondly, construct dimensions were identified and specified through the literature review and theoretical considerations to measure causes and effects. Initially, variables were categorized according to interrelationship, such as type of variable – latent or manifest –, scale, and procedure to select variables for this research.

This study aims at establishing a substantial empirical theory by bringing theoretical and observed variables into a relationship with correspondence rules (Bagozzi, 1998). A causal analysis and modeling approach was selected to evaluate the data as it offers stricter testing of theory compared to bi- or multivariate statistical methods to test relationships (Fassott & Eggert, 2005; Hildebrandt & Homburg, 1998). The main advantage of causal analysis is considering measurement error in the surveyed variables through differentiation between latent and indicator variables (Albers & Hildebrandt, 2006; Homburg & Dobratz, 1998). The latent construct domain was specified, and a sample of items was generated from the literature review. The resulting hypothetical measurement is the interrelation amongst relevant manifest and corresponding latent variable (Bagozzi & Yi, 2011). Subsequently, selected latent variables embody or capture variance detected in manifest variables.

A critical examination of existing measurement scales was conducted before integration into the study. The applied scales are either new or modified to fit the context of negotiations or fully adopted existing scales in their original form. Several items for the respective constructs were collected and evaluated to see if these explain the construct's content coherently. All constructs were operationalized via a multi-item measure of at least three items (Bergkvist & Rossiter, 2007; Churchill, 1979). As a result, it is essential to confirm that all operated indicators are attributes of the intellectual and contoured phenomenon (Edwards & Bagozzi, 2000). The number of items was reduced, according to Drolet & Morisson (2001), wherever possible, as the information content after the third item is marginal and only sometimes improves the measurement quality. This method eliminates multi-item abuse in which items are adjusted until inner consistency, conforming with Cronbach Alpha thresholds, is achieved. The content structure of items was questioned to maintain this limited item approach. This conduct avoids conceptual errors and wrongful statements disguised by a demanding method. With this approach, negotiation experts were not unnecessarily engaged, and other intriguing questions could be placed in the survey. In addition, it was tested if items are of reflective or formative nature using the question catalog by Fassot & Eggert (2005) built on work of Jarvis et al. (2003). This question catalog classifies all items as reflective. Moreover, constructs that measure success – negotiation effectiveness and efficiency – are output values and thus are always reflective. Henceforward, an abstract reflective formulation of indicators is used for the theoretical test to precisely mirror attributes of a latent variable (Herrmann et al., 2006). Subsequently, a content validity verification with 30 experts was undertaken to evaluate measurement indicators for construct dimensions. As a multi-item questionnaire (Churchill, 1979), the respective reflective constructs are measured with a five-point Likert Scale employing closed questions (Likert, 1932). Default answering categories characterize this questioning type by simplifying the evaluation (Raab-Steiner & Benesch, 2015). The following chapter evaluates measurement reliability and validity of constructs to develop norms.

As a management-oriented research study with decision relevance, a good explanation of variance and prediction of target variables is essential (Herrmann et al., 2006). Data for this study was collected through subjective assessment of negotiators and examined statistically via a variance-based structural equation model for behavioral and managerial sciences (Bagozzi & Yi, 2011). This approach assesses the novel theory-based hypotheses structure between latent constructs (Herrmann et al., 2006) and targets transparence on variance in two focus variables to identify the central determinants. The needful causal modeling in a structural equation model was conducted in SmartPLS 4, developed in the original version by Apel in 1977 and adapted by Lohmüller in 1982. An appraisal of models – measurement and structural – was undertaken for empirical examination of the causal model. Further, the results of the survey were compared to the hypotheses. After the evaluation of hypotheses, the relative meaning of the examined effect sizes was evaluated. The theory is successfully fulfilled if the whole model can reproduce its data sufficiently. The structural equation model was then tested for moderating effects by analyzing the interaction term (Hayduk, 1987) as the moderator variables were measured via a continuous scale level (Jaccard & Wan, 1996). In Chapter 4, formulated hypotheses are examined, and the results are condensed according to relevance and interpreted.

4 Analysis & Results

4.1 Quantitative Data Generation

Empirical data generated through this quantitative analysis identifies three statistical evaluations presented in the following sections: 1) descriptive statistics, 2) explorative statistical dependencies as well as 3) effect mechanisms in a model of structural equations. Identified directions and strengths of the effect-mechanisms in the structural equations are compared with the prior formulated theories. The sampling method used is a nonprobability sampling, which includes convenience paired with judgment and snowball sampling to minimize bias (Ben-Shlomo et al., 2013). Participants were approached through a comprehensive online search from September 2022 to May 2023. The gross sample consisted of 2521 clicks on the survey link, of which 918 participants tried to complete the survey (35.90%). A total of 185 participants (20%) of 918 were terminated from the survey as they did not participate in a negotiation in the last 12 months. From this, 733 (79.84%) questionnaires were used to test the hypotheses. In sum, 376 participants (29,35%) interrupted the survey before the completion process. The consent form was rejected by 19 participants (1.41%), which is a neutral loss to the overall sample. Two cases were discarded due to a respondent's misconduct. The average processing time was 14 minutes with the most access at 6 pm of 198 participants accessing the survey during this time. The typical participation rate daily amounts to 14 participants for this study. At the same time, the mediocre figure for participants per week sums up to 90 individuals. The survey question with the most terminations requested information on the percentage share of negotiation channels used with a total of 60 survey terminations. The sample consists of international negotiation managers from diverse companies and industrial sectors who were requested to place themselves in recently completed negotiation situations to use retrospective data as an analysis unit.

A confirmatory or explanatory factor analysis was not piloted due to the adequate theoretical basis, the homogeneity of sampling, and the heterogenous measurement. If there is a well-established theoretical basis or prior empirical evidence to suggest the core factor configuration of a measure, then an exploratory factor investigation may not be necessary (Bandalos & Boehm-Kaufman, 2010). Furthermore, the sample is not homogeneous regarding relevant markers such as age, gender, culture, or socioeconomic status. In that case, a confirmatory factor analysis is not appropriate (Marsh & Hau, 1996) because different groups may have different factor structures, which would lead to invalid or biased results. At last, if the measure is very complex, with many items or multiple scales, then a confirmatory factor analysis may be difficult or impractical to conduct (Gorsuch, 1997).

4.1.1 The Demographic Portrait of Negotiators

Every statistical analysis begins with a descriptive data exploration. The statistical descriptive section is the base for the explorative and inductive statistics parts. It commences with the presentation and description of frequencies in characteristic values. Most diagrams are generated from frequency tables which enable the identification of correlations. With this critical analysis, initial impressions are gained as a basis for further analysis.

Most participants in this investigation are male (72%), whereas 24% of females participated in this research. Once the gender was selected as diverse, while 3.1% of participants preferred not to disclose their gender. The average mean participant's age is 31 years. The median age of negotiators, within the scope of this study, is 42 years. The age categories are relatively evenly distributed, with 60 negotiators equally in the category below 29 years and above 63 years. The age category 30-39 years holds 187 participants, and the age group 50-59 years contains 156 negotiators. Whereas the age grouping 40-49 years encompasses almost 200 participants and represents the largest age category of survey respondents.

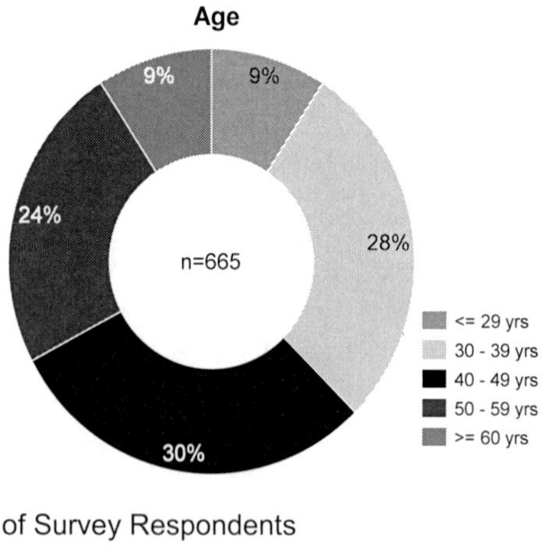

Figure 5. Age Groups of Survey Respondents

Over 60% of survey participants (394 counts) hold European nationalities, followed by 13% Asian & Asian Pacific or South & Southeast Asian nationalities (82 counts). A meaningful sample size originates from North America (53 counts), and 27 respondents hold South American nationalities. African negotiators are of a substantial sample size with 37 participants, while respondents with a Middle Eastern nationality account for 21 respondents. The study participants from Europe can be divided into 37 Eastern, 40 Northern, 17 Southern, and 292 Western Europeans. The group of Western Europeans consists of a large group of German nationals (257 counts).

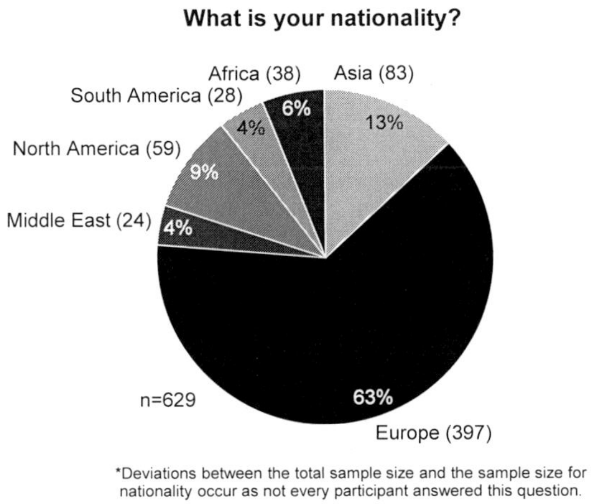

Figure 6. Nationality of Survey Respondents

One-half of the negotiators work in small-sized companies with fewer than 50 employees (228 counts) or 50-250 employees (128 counts). Evenly distributed, more than 10% of respondents work in medium-sized enterprises consisting of either 251 to 1.000 employees (13.5%, 99 counts) or sized from 1.001 – 5.000 employees (11.7%, 86 counts). The smallest group of respondents (5.2%, 38 counts) work in companies with a staff number ranging from 15.000 to 50.000 or from 5.001 to 15.000 with 62 counts (8.5%). A total of 12% of respondents (88 counts) work in large organizations with over 50.000 employees.

A third of the representation accounts for experts on industrial goods (5%, 36 counts), negotiators in the automotive industry (14.5%, 106 counts), or high technology industry employees (10%, 76 counts). Around 50 respondents are engaged in financial services (7%, 51 counts) and the public sector (6.5%, 48 counts). Respondents from the chemical, petrochemical, and pharmaceutical sectors comprise 7% of the sample. A minority of 28 participants in this study work in media and telecommunications. Participants are also engaged in the following other industries, which were not explicitly listed as answer categories in the survey: construction, defense, training & coaching, information technology, legal, and consulting (264 counts, 36%).

Most survey respondents negotiate in the functional area of management (35.7%, 262 counts). The second highest represented functional area is procurement (15.3%, 112 counts). The least number of participants work in the field of mergers and acquisitions or finance (.8%, 6 counts).

A total of 74% of respondents (545 counts) are performing in a management position. Only 76 respondents (10%) are team members, assistants, or administrators. Managing directors (25%, 185 counts) and managers without a team (15%, 111 counts) are strongly represented in the sample. The second most important category is divisional manager (12%, 88) and department head (11%, 83). Under the category other position, the following professional positions were named: consulting, legal, academia, owner, self-employed, and even head of a union. Almost 60% of survey respondents negotiate in Europe, followed by the most prominent negotiating operations by size in North America (14%) and Asia (12%).

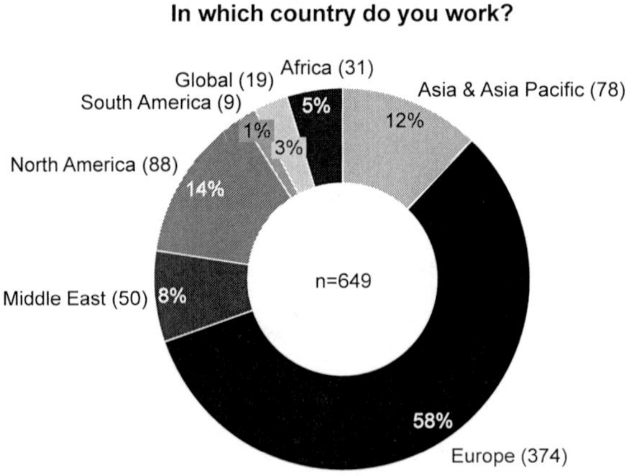

Figure 7. Geographic Diversity in Global Negotiator Operations

It is assessed if the negotiators of this study have prior negotiation experience towards the negotiation situation, the negotiation object, or the negotiation partner (Carlson & Zmud, 1999; Fulk et al., 1990; Plinke, 1985; Voeth & Herbst, 2015; Walther, 1992; Woodside & Davenport, 1974). The attitude and behavior towards media depend on the, through professional tasks, developed work experience. Almost all participants of this study have prior experience in negotiations (95.5%, 693 counts) with an average negotiation experience of 15 years.

Multiple and continuous interaction touchpoints between negotiation parties enhance negotiation partnerships (Holmlund-Rytkönen & Strandvik, 2005). Therefore, it is vital to examine if negotiators possess a mutual negotiation history and how this impacts the negotiation process and outcome (Carlson & Zmud, 1999; Fulk et al., 1990; Li, 1996; Walther, 1992; Parsons & Wooldridge, 2002; Zhang et al., 2004). Within this study, it is analyzed if an effect is intensified through the existence of a negotiation history between the negotiation partners. A cooperative behavior can be shaped by a mutual history or a positive evaluation of the receiving end. More than half (54%) of this study's negotiators (396 counts) possess a mutual negotiation history with their negotiation partner. On average, a mutual negotiation history between negotiators within this study accounts to seven years. This confirms the trend towards the formation of sustainable negotiation partnerships (Eklinder-Frick & Åge, 2020).

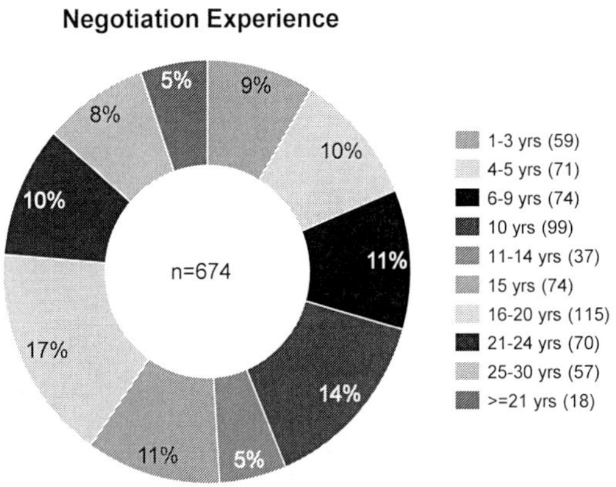

Figure 8. Negotiation Experience of Survey Respondents

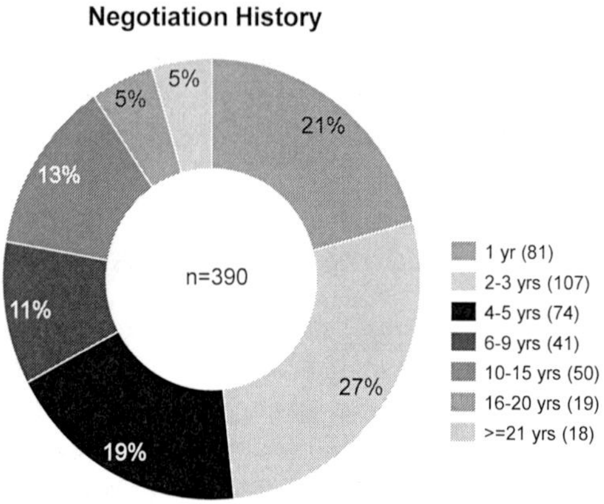

Figure 9. Negotiation History of Survey Respondents

4.1.2 Quantity of Negotiation Channels Applied

At least two negotiation channels were used during the most recent, relevant negotiation by 702 negotiators (95.8%). Only 31 negotiators (4.2%) used a single channel to negotiate. More than half of survey participants use two to three negotiation channels, predominantly classic negotiation channels of email, face-to-face, video, and phone calls. The trend of using two to three negotiation channels is closely followed by using four negotiation channels with 156 counts of total respondents. In fact, 75% of respondents (539 counts) use two to four negotiation channels (see Figure 10). Additionally, 6-8 channels are the least chosen.

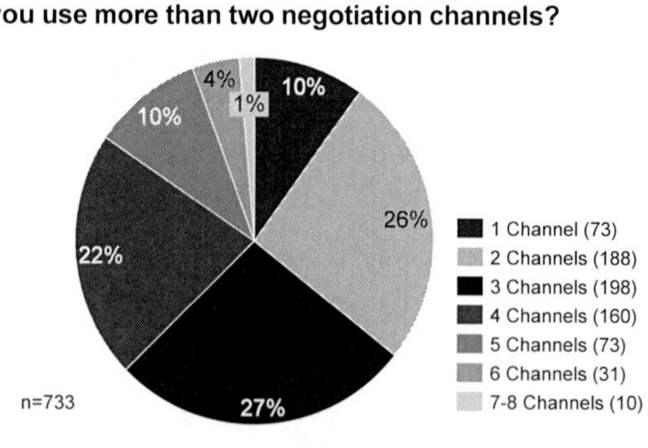

Figure 10. Channel Quantities in Negotiations

Men apply slightly more negotiation channels (3.2 mean value) than females (2.9 mean value) see results in Annex 9. However, this discrepancy is insignificant (Levene Test for Variance Equality p-value .227). The least channels are utilized by the youngest (2.7 mean value) and oldest negotiators (2.9 mean value). Most channels are used by negotiators aged 30-39 years (3.3 mean value) and 50-59 years (3.3 mean value). A significant correlation exists between age and channel quantity in negotiations (Pearson-Correlation p-value .014), which signifies that the older the negotiator, the more channels are employed with an overall weak effect (Pearson Correlation .091).

On the one hand, if negotiation experience is prevalent, then negotiators tend to apply approximately one additional channel (3.1 mean value) than if a negotiator has no prior negotiation experience (2.2 mean value). The least negotiation channels (2.5 mean value) are used by negotiators with the least experience (1-3 years). Most negotiation channels (3.5 mean value) are utilized by negotiators with the most experience (25-30 years). The influence of experience in bargaining on the number of negotiation channels employed is highly significant (Pearson-Correlation <.001), which implies that the more experienced a negotiator is, the more channels are utilized (Pearson Correlation .166 translates into a weak effect). On the other hand, a long negotiation history equates into a high number of channels used (3.5 mean value). The impact of a prevailing negotiation history with the negotiation partner on the quantity of negotiation channels is insignificant (Pearson-Correlation .109). Hence, there is no correlation between the negotiation history with the specific negotiation partner and the application of channels (Pearson Correlation .059).

4.1.3 Channel Choice in Negotiations

With multiple response choices, participants selected the negotiation channels used during their most recent, significant negotiation (see Figure 11). Evidently, the most prominent negotiation channel employed is email with 491 counts (68%), followed by negotiators selecting 64% face-to-face (464 counts) and 62% video conferencing (451 counts) during their last relevant negotiation. Additionally, 26 respondents quote that they use alternative channels for their negotiations, even in the traditional form of letters specifically for exchanging formal documentation or even texting as a timeless alternative for negotiators. As an online document editor, Google Docs is a stable alternative, especially for internal negotiations or bargaining with trusted negotiation parties. Alternatively, an online human resources portal is the internal negotiation solution for respondents working in human resource management. Physical auctions and digital auction platforms are valued alternatives for purchasing experts, such as web bidding platforms or E-sourcing like SAP Ariba accounts management software are referred to as trusted channels as well as eBay, the online auction and shopping website, and Vinted (an online platform for pre-owned goods). Ultimately, LinkedIn – the world's largest professional network on the internet – is more frequently used by negotiators for heterogenous negotiations. Finally, social media postings are also preferred by some negotiators.

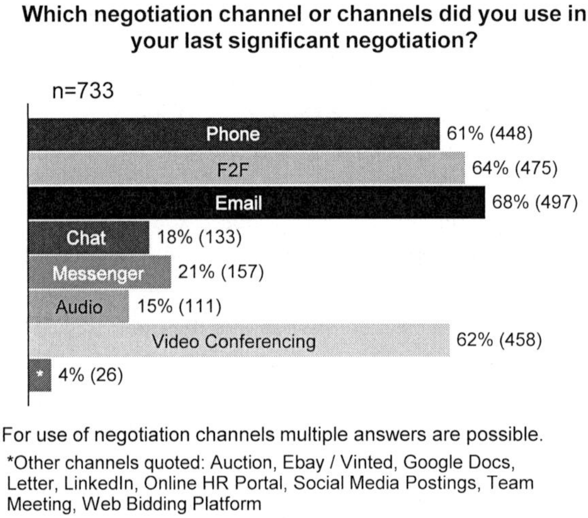

Figure 11. Channel Choice in Negotiations

The weight of age on the media choice phone call, face-to-face, email, chat, messaging, audio conferencing, and other channels is insignificant. Despite this insignificant result, there is a highly significant relationship (Levene-Test for Variance Equality p-value .000) between the negotiator's age and the selection of the channel video conferencing. As a result, older negotiators tend to use more video conferencing in their negotiations than younger negotiators. A T-Test for Average Mean Equality also identifies this age and video conferencing relationship as significant (p-value .013) with a minor effect strength (Cohen's d .201).

The bond between more experienced negotiators and utilizing several channels is insignificant (for details, see Annex 19). However, negotiators with more experience use phone calls significantly more than those with less experience (Levene's Test p-value .001, T-Test for Equality of Means p-value .007 for variances equality not expected; with a weak independent sample's effect size Cohen's d .220). More experienced negotiators use email more frequently (mean 11.57, std. deviation 22.936) than negotiators with less experience (mean 3.23, std. deviation 34.702) with the Levene's Test yielding significant results (<.001 for equal variances assumed). Accordingly, the T-Test for Equality of Means confirms its significance <.001 for equal variances not assumed. Overall, the strength of these measured differences is weak (Cohen's d .306). More experienced negotiators use video conferencing more often (mean 11.34, std. deviation 22.929) than those having less than half of the experience (mean 4.79, std. deviation 33.493) with the Levene's Test yielding significant results (<.001 for equal variances assumed). A T-Test for Equality of Means confirms its significance .004 for equal variances not assumed. Overall, the strength of these measured differences is weak (Cohen's d .240).

4.2 Channels Determine the Process Outcome

4.2.1 Drivers for Selecting Negotiation Channels on Purpose

Overall, 67.4% of participants (494 counts) in this study select a negotiation channel on purpose and mostly reveal why they do so. Only 239 participants (32.6%) stated they did not intentionally select a negotiation channel. To deep dive into why negotiators select negotiation channels on purpose, a structured qualitative content analysis (Forman & Damschroder, 2007; Lindgren et al., 2020; Mayring, 2004; Selvi, 2019; Vaismoradi & Snelgrove, 2019) promises insights by evaluating respondent's survey comments. “When data allow the interpretation of latent content, qualitative content analysis can reveal the depth and meaning” of respondents' textual comments (Lindgren et al., 2020, p. 6). Hence, the following research question: “In general, do you select specific negotiation channels on purpose for your negotiation? If yes, explain why” becomes the basis for the qualitative content analysis. The assessment of free-text comments aims to accumulate new theoretical observations regarding this research question. A coding guideline defines the criteria for evaluation of the material (see Table 4). The qualitative content examination complies with the quality criteria of qualitative research by making the procedure transparent on how measurement has taken place, while ensuring its reproducibility and excluding considerable subjectivity. To warrant reliability, the same coding guideline and category system was applied to the whole material derived from this survey.

A deductive as well as inductive process establishes categories and by assigning text passages of respondents to categories, the coding guideline structures content systematically. “Abstraction and interpretation during the analysis process” are descriptive and concrete by creating selected and condensed meaning units (Lindgren et al., 2020, p. 4). Some categories were prevalent before coding, and others emerged during the coding process while sighting the free text responses of survey respondents. A definition of a category with an anchor example quoting concrete text supported in assigning content to categories (Mayring, 2015). For inductive category building, on the one hand, paraphrasing and generalizing was one approach; on the other hand, incorporating integral elements of the original qualitative material as citations from open questions was another approach. In case of demarcation problems between categories, according to Mayring (2015) the rules formulated enable a transparent allocation. Based on the coding guideline, the respondent’s free text was assigned to a matching category to show why the selection of negotiation channels happens on purpose. A software program such as Maxqda, used for qualitative content analysis, was not applied since only a single free text question occurs in this survey. The free text was inserted into the table of the coding guideline and placed in its corresponding category. This process results in an overview of all relevant statements of respondents for one category. Accordingly, this overview summarizes the results regarding this survey question and answers the research question of why a selection of negotiation channels happens on purpose.

Table 4. Coding Guideline for Selecting Channels on Purpose

Category	Definition	Anchor Example	Coding Rule
Process Outcome: Negotiation Context	All text passages which indicate that the respondent of the survey refers to the process outcome and specifically the negotiation context as a reason to select negotiation channels on purpose.	Respondent #526: "Depending on the context and the sensitivity of the negotiation."	Text must contain defining aspects of the "negotiation context."
Process Outcome: Negotiation Management	All text passages which indicate that the respondent of the survey refers to the process outcome and specifically the negotiation management as a reason to select negotiation channels on purpose.	Respondent #250: "Depending on the purpose and strategy as well on the result I want to achieve."	Text must contain defining aspects of "negotiation management."
Process Outcome: Negotiation Behavior	All text passages which indicate that the respondent of the survey refers to the process outcome and specifically the negotiation behavior as a reason to select negotiation channels on purpose.	Respondent #56: "I prefer face-to-face or video conferences as I can bring in my personality and can read the counterpart's body language."	Text must contain defining aspects of "negotiation behavior."
Economic Outcome	All text passages which indicate that the respondent of the survey refers to the economic outcome as a reason to select negotiation channels on purpose.	Respondent #503: "Face-to-face provides a more intimate setting for authentic connection resulting in win-win negotiation."	Text must contain defining aspects of the "economic outcome" of a negotiation.
Socio-Emotional Outcome	All text passages which indicate that the respondent of the survey refers to a socio-emotional outcome as a reason to select negotiation channels on purpose.	Respondent #917: "I prefer personal contact (face-to-face) because you can build up a different level of relationship and also experience facial expressions and gestures in the course of the negotiation and classify them accordingly."	Text must contain defining aspects of the "socio-emotional outcome" of a negotiation.

Most negotiators (95%, 696 respondents) shared comments on why they select negotiation channels on purpose. Textual answers of survey respondents can be divided into three categories for selecting a bargaining channel on purpose to influence the 1) process outcome, 2) economic outcome, or 3) socio-emotional outcome of the negotiation. While stating the reasons for selecting negotiation channels on purpose, study participants named one or multiple categories with corresponding subcategories. The most stated reason for 50% of respondents (351 counts) is to select negotiation channels to influence the bargaining process outcome; 11% of negotiators desire to impact the economic result (75), and 9% (60 counts) want to influence the socio-emotional outcome. The reasons named for influencing the process outcome can be divided into the following subcategories: 1) negotiation context, 2) negotiation management, and 3) negotiation behavior to which keywords are assigned in Table 5.

Scanning through 351 process outcome keywords, 92 respondents (26%) identified the negotiation context as a critical reason to select negotiation channels on purpose. Most prominently, the keyword negotiation object is mentioned on 32 counts, followed by the term negotiation partner (18 counts) as a reason to select a negotiation channel on purpose. Subsequently, 200 respondents (57%) named negotiation management a critical reason to choose negotiation channels on purpose. The essential keywords identified in the subcategory of negotiation management are 1) strategy & tactics (72 counts), 2) documentation (40 counts), 3) time management (31 counts), and 4) convenience (18 counts) see Table 6.

Table 5. Process Outcome Subcategories and Keywords

Category	**Subcategory**	**Keywords**
Process Outcome	Negotiation Context	Negotiation partner Negotiation objects Negotiation media Negotiation communication Negotiation stage, process, phase Number of negotiations External factors Culture Language
	Negotiation Management	Situational analysis Stakeholder preparation Strategy, tactics Formal vs. informal Steering, control, screening, structure, precision Revision, rehearsability, review, reflection Level of information sharing, visual information, data sharing, data clarity, clarification, data management, document sharing Time management Geographic distance, location, travel time / zone / restriction Cost efficient Convenience, comfort, confidence, intimacy, simplicity, flexibility Alignment, acceptance, agreement, mutual understanding Transparency, confidentiality, security Documentation, avoidance of documentation Company policy determines channel, channels dictated by public procurement law
	Negotiation Behavior	Body language Nonverbal cues Reactions Facial expressions Emotions Feeling Gestures Verbal cues Eye contact Empathize

Table 6. Keywords for the Subcategory Negotiation Management

Rank	Keyword	Count
1	Strategy & Tactics	72
2	Documentation	40
3	Time Management	31
4	Convenience	18

Conclusively, 93 respondents named negotiation behavior as a critical reason to select bargaining channels on purpose to influence the process outcome. This discovery supports the study of Voeth & Becker (2013) that negotiators analyze the negotiation partner with regard to the expected negotiation behavior including the opponent's negotiation performance, strengths, weaknesses and preferences. The most prominent reasons for selecting a channel are the ability to see and interpret the opponent's body language, nonverbal cues, reactions, facial expressions, and emotions as depicted in Table 7.

Table 7. Keywords for Behavior as Driver for Channel Choice

Rank	Keyword	Count
1	Body Language	33
2	Nonverbal Cues	14
3	Reactions	14
4	Facial Expressions	12

As 11% of respondents (75 counts) choose negotiation channels to influence the economic outcome (see Table 8), a deep dive clarifies the underlying root causes for this objective. Specifically, for a better economic outcome, channel choice supports shaping the final agreement formulation, mutual result(s), or outcome(s), the enhancement of individual performance, joint profit(s), and effectiveness as well as efficiency of the overall negotiation. Respondent #140 uses channels "to increase the chance for successful negotiations." Most negotiators within this category mention individual performance, effectiveness, and efficiency as subcategories. Negotiator #120 claims "face-to-face has proven most effective," and negotiator #1198[1] wants "to lead the agenda and select the channel" as "this can give a significant advantage over the other party and increases certainty."

Only 59 of 696 study participants (8.5%) selected negotiation channels on purpose to enhance the socio-emotional impact of bargaining. Establishing a relationship and trust are prominent reasons for selecting a negotiation channel within the socio-emotional category (see Table 9). The negotiator #591 wants to "build a relationship," while for the negotiator #684, channel choice "depends on the relationship," for respondent #293, channel choice is "specific to the relationship I have built with the other party." Respondent #408 thinks about "how I can create trust, transport my interests, and discover my counterpart's interests." Relationship and trust are two interdependent themes in negotiations - one occurs with the other and vice versa, while trust can mean different things depending on the relationship (Ross & LaCroix, 1996).

Results show that most negotiators in this study select one or multiple channels actively to influence the process outcome of the negotiation. These findings, retrieved from 696 international negotiators, confirm the strong impact of channels on the bargaining procedure (Bazerman et al., 2000; Galin et al., 2007; Geiger, 2020; McGinn & Croson, 2004; Poole et al., 1992). In detail, negotiators in this study report how to steer the negotiation process actively by using negotiation channels on purpose. The next section discusses findings regarding the specific channel(s) used by international negotiators during the negotiation stages.

[1] In this specific case, the numbering of the respondent is unusually high at #1198 as the survey software also counts respondents into the total amount that terminated the survey or were not eligible for participation due to the lack of a relevant negotiation within the past 12 months.

Table 8. Channel Selection to Influence the Economic Outcome

Category	Subcategory	#	Negotiator Survey Comment
Economic Outcome	Results	62	*I prefer to negotiate face to face as this enables me to get better results for my decisions*
		140	*To increase the chance for successful negotiations*
		250	*Depending on the purpose and strategy as well on the result I want to achieve*
		702	*Generally, to try utilizing best option for the output desired but sometimes dictated by circumstances, counterpart, etc.*
	Individual Performance	131	*Face-to-face or video conference allows me to rather be more convincing (video/face/personal) as well as nice/gentle/determined tone in emailing*
		174	*I can choose the channel I like, the easiest ones for me especially*
		516	*The channel influences my communication abilities, face to face negotiation provides the best setting, then video conferencing*
		1045	*Easier to explain the difference, to present my point of view*
		1198	*It is important to lead the agenda and select the channel, this can give a significant advantage over the other party and increases certainty.*
		1322	*I perform well in videoconferencing.*
		1362	*Zoom - cause it's easier to share screen for presentations, negotiations and getting the job done.*
	Joint Profit	503	*Face to face provides a more intimate setting for authentic connection resulting in win-win negotiation*
		705	*1:1 phone call is preferred once data clarity is given. In 1:1 dialogue my experience it that it is the best way to find a mutually acceptable solution (win/win)*
		752	*To set the criteria of both parties' requirements on the table and negotiate the win-win for solutions.*
	Effectiveness	120	*Face to face as it has proven most effective for me*
		248	*Phone call is most effective*
		898	*Face to face is more effective*
		1115	*Depending on cases, phone is effective*
		1184	*I prefer the channel considering the most effective time to conclude.*
		1347	*It is the most effective (face to face) by .... far*
	Efficiency	268	*Time-management and efficiency*
		949	*Via phone/video chat or in person is more efficient and expedient*
		1075	*F2F (and during corona times video) is most efficient when it comes to decision making*
		1119	*Efficiency and speed to get to a result/conclusion*

Table 9. Relationship & Trust as Drivers for Channel Selection

Category	Subcategory	#	Negotiator Survey Comment
Socio-Emotional Outcome	Trust & Relationship	213	*Face-to-face to build trust and relationship (especially in Asia)*
		281	*Start with face-to-face to establish rapport, follow-up can be mail or phone*
		293	*Specific to the relationship I have built with the other party, I choose what I feel most comfortable with*
		312	*Depends on level of closeness to person - e.g., very close is face-to-face. Acquaintances would be more digital. Also depends on impact of the negotiation on the human relationship with the person. Again, if the relationship is very important then face-to-face.*
		558	*F2F for building a relationship, virtual setup for focusing on content, chat for informal exchange of ideas and input*
		591	*F2F is more effective & vendor or supplier are more obliged when they visit our premises*
			I'm trying to build a relationship, so I want as many clues as possible. 1st choice, in-person at their facility. 2nd choice, at my facility or neutral. 3rd choice video. 4th choice phone. Chat message is very personal, a different clue.
		622	*Face to face allows negotiation in relationships (distribution channels and supplier relations)*
		676	*If I need to establish a relationship or use my interpersonal skills, I prefer in-person. For speed and convenience, I use video or phone. For asynchronous communication, working out details and to document the negotiation, I use email.*
		684	*Depends on the relationship, the better the relationship (or the longer, the less face to face comes into play*
		760	*If it's possible then yes, it's not always up to me. But if I do, I choose channels that allow for a live conversation (video/face-to- face). It's easier to build rapport and read the audience this way.*
		917	*I prefer personal contact (face-to-face) because you can build up a different level of relationship and experience facial expressions and gestures during the negotiation and classify them accordingly.*
		925	*Depending on the maturity of the relationship with the counterpart and on the specific phase within the negotiation*
	Trust	287	*Body language and persona appearance defines trust*
		392	*Face-to-face to foster trust through embodied interpersonal contact*
		408	*I think about how I can create trust and how I can transport my own interests and find out my counterpart's interests best.*
		691	*Face to face, all ingredients in 1 place n more fast, trusted n accurate*
		1209	*the selected channels depend on which stage the negotiation is in. Initial contact can be done via email, but more detailed and in-depth conversations need to be done in person to influence and to build trust with the other party*
		1213	*I believe negotiations are quicker and the result more long lived when face to face or video chatting. It is easier to establish trust and ensure alignment*

4.2.2 Identified Trends in Channel Use Along Negotiation Stages

The analysis of negotiation stages is an essential aspect for comprehension of multi-channel negotiations. Two stages in a consensus building process were initially defined by researchers in the distinction of formula & detail (Ikle, 1964) or differentiation & integration. Graham (1987) progresses further and divides the negotiation into planning & preparation, interaction, and a closure phase. As Weber (2019) depicts in her dissertation, a negotiation consists of a pre-negotiation phase, a transition phase for the offer introduction and a main negotiation phase – the latter being repetitive in several rounds. In accordance, survey questions 8 to 12 explore the use of negotiation channels along the various negotiation stages by requesting negotiators to revert to their last significant negotiation, which must be closed and completed for the study. For this exploration, channels were assigned to the subsequent negotiation stages: (1) greeting & initiation, (2) background information exchange, (3) agenda setting, (4) issue discussion, and (5) final agreement formulation. Derived from Alfredson & Cungu (2008), this stage definition originates from Zartman & Berman's (1982) diagnostic, formula, and details phase in negotiations and studies of Yuan et al. (2003). Overall, 722 valid responses were obtained for this analysis. In line with Weber (2019), awareness prevails that essential negotiation conduct with a pivotal impact does occur outside these formalized negotiation processes. It needs to be considered that according to Lenzing (2015), negotiation management does not always formally end with the closure phase and can lead into renegotiations (Salacuse, 2001; Schoop et al., 2008).

All three negotiation stages from greeting & initiation, background information exchange, and agenda setting exhibit a strong email tendency. This aligns with outcomes of Voeth et al. (2022), that negotiators see the largest benefit in electronic support for the negotiation preparation. Data strongly supports that negotiation stage (1) greeting and initiation is conducted via email with 321 counts. The negotiation stage (2) for background information exchange is also intensely conducted via email with 400 counts. The negotiation stage (3) agenda-setting is also predominantly conducted via email.

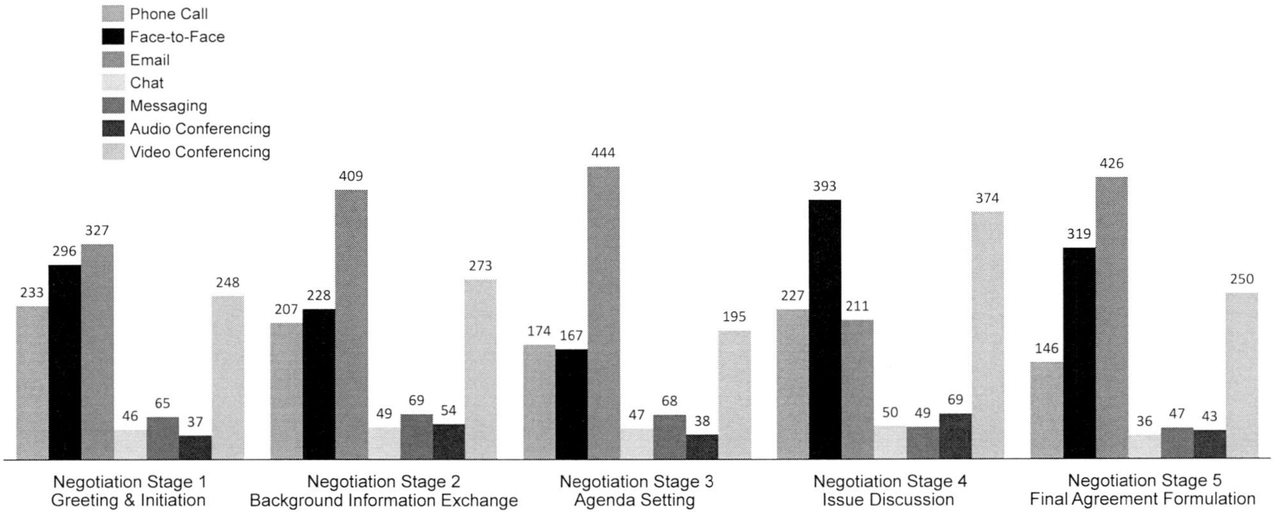

Figure 12. Media Use Along the Negotiation Stages

Conclusively, agenda setting is not necessarily done via face-to-face meetings (164 counts). Data suggests that the negotiation stage (4) issue discussion is performed via face-to-face or video conferencing. It is visible that an issue discussion is not necessarily done via email with 207 counts. It aligns with Eidhoff (2017) that digital technologies should not be induced in every process phase. The negotiation stage (5) final agreement formulation is prominently conducted via email (419), followed by face-to-face (315) and video conferencing (247). It may be concluded that the media channels chat, messaging, and audio conferences play a minor role concerning specific negotiation stages. With the negotiation medium chat being the least essential channel in negotiation stage planning, followed by audio conferencing and messaging. A phone call is still essential for negotiation stage (1) greeting and initiation, as well as negotiation stage (4) issue discussion. The importance of a bargaining phone call decreases in negotiation stage (5) for a final agreement formulation.

On 59 occasions (8%), the use of an alternative negotiation channel along the negotiation stages was stated by the study participants in Table 10. Accordingly, officially written, and sealed documents – in the form of legal court protocols or hard copy formal letters – are essential to close the deal in negotiation stage (5) the final agreement formulation. At some point, file sharing is necessary to negotiators by making a distinction between internal and external file sharing. Pre-bargaining research is essential through search-specific media like Wiki in negotiation stages (2) and (3). Multi-mode communication like Microsoft Teams is used in the first and second negotiation stage and use of Zoom in all negotiation stages is regarded as mandatory. Subject to the contextual form, type and setting of the negotiation, third parties can be introduced for a background information exchange, according to respondents. Depending on the type of negotiation, specialized industry-specific platforms are used to seal the deal.

Table 10. Alternative Channels Applied within Negotiation Stages

Alternative Channels	**Negotiation Stages**					
	1	**2**	**3**	**4**	**5**	
Legal Court Protocol					x	
Soft Copy Documents					x	
Hard Copy Letter	x				x	
Texting		x		x		
SharePoint		x				
Google Docs					x	
Microsoft Teams	x	x				
Zoom	x	x	x	x	x	
Electronic Signature System					x	
LinkedIn	x			x		
Social Media		x		x		
Instagram Messaging	x					
Government Portal		x	x	x	x	
Supplier Platform		x	x		x	
Procurement or Sourcing Platform	x	x	x	x	x	
Auction Platform		x			x	
Wiki		x	x			
Internet Based Research		x				
Artificial Intelligence			x			
Third Party Legal Counsels or Attorneys		x				
Talks with Employees		x				
Business & Industry Forums & Events	x					
Total Alternative Channel Counts	12	15	10	6	16	59 (8%)
Negotiation Stage is not Required	12	19	46	5	5	87 (12%)

4.3 Validating Synchronicity Theory

Considering Synchronicity Theory to measure the effectiveness of various communication channels to reach conveyance (information sharing) and convergence (decision making), participants responded to the question of ranking favorable media used to exchange information or to achieve a consensus. Negotiations are characterized by an information exchange and decision-making process (Schwarz, 2012). Information sharing is the exchange of facts and background information that is helpful to the course of the negotiation. The amount of needful information to counteract wrongful decisions and to attain effective and efficient negotiation results is subjective to the individual negotiator (Schwarz, 2012). According to Schwarz (2012), information augments the decision quality as a central steering component in the analysis and steering of negotiation performance. Highly significant differences (Friedman-Test p-value <.001) occur in ranking the importance of media channels for information sharing. According to participants of this study, the most crucial negotiation channel for sharing information with the negotiation party is the medium of face-to-face (1.76 mean value), followed by video conferencing (2.20 mean value) and email (2.41 mean value). These three channels are rated on average on rank two in importance. Video or audio conferencing, email, and phone calls are close at a mean rank of around 2.2 to 2.8. This result signifies that all three channels are equally well perceived for information sharing. Messaging (3.46 mean value) and chat (3.35 mean value) are the least helpful information-sharing channels.

Table 11. Ranking of Negotiation Channels for Information Sharing

Rank	Channel	Mean Value
1	Face-to-Face	1.76
2	Video Conferencing	2.20
3	Email	2.41
4	Phone Call	2.65
5	Audio Conferencing	2.84
6	Chat	3.35
7	Messaging	3.46

This also applies to consensus building, whereby highly significant differences (Friedman-Test p-value <.001) in the assessment of the importance of media channels occur in Table 12. For consensus-building, the most important channels are face-to-face (3.22 mean value), video conferencing (3.60 mean value), and phone calls (3.83 mean value). The main difference to information sharing is that respondents opt for a phone call instead of email to arrive at a mutual decision. Subsequently, it can be confirmed that all three channels display audio functions which are inherently crucial for decision-making. Opposed to the results for information sharing, the channels for decision-making do not lie close together in ranking and are therefore not interchangeable. In line with information sharing, chat (4.51 mean value) and messaging (4.61 mean value) are the least useful for consensus building. In addition, the Mann-Whitney Test proves the insignificance of gender on the ranking of negotiation channels for information sharing or consensus building.

Table 12. Ranking of Negotiation Channels for Consensus Building

Rank	Channel	Mean Value
1	Face-to-Face	1.46
2	Video Conferencing	2.02
3	Phone Call	2.33
4	Audio Conferencing	2.61
5	Email	2.78
6	Chat	3.25
7	Messaging	3.31

4.4 Explorative Dependency Analysis

As a first step, it is tested if the following variables age, gender, nationality, negotiation experience, negotiation history, functional area, position, industry, organizational size, and country of operation influence the 1) selection of a specific type of negotiation channel. The second step analyzes if the variables impact the 2) quantity of negotiation channels selected. For this investigation, various statistical methods were deployed depending on the variable type and measurement level. Previous Sections 4.1.2 and 4.1.3 highlight age and experience's significant influence on the channel choice and the number of channels employed. However, a prevalent negotiation history does not affect motivations for channel choice, or the number of channels employed. The subsequent explorative dependency analysis shows an overall impact of gender on channel choice and its specific effect on the preference for a negotiation channel. The negotiators' industry and functional area of occupation influence channel choice, while the professional position of the respondent even simultaneously impacts both the channel selection and the quantity of media used. The organizational size also influences channel choice. However, no correlation exists between the size of the company and the employment of several negotiation channels. The nationality and country of operation impact channel choice substantially yet there is no significance between the number of channels selected. The results are discussed in detail in the succeeding sections.

4.4.1 Industry & Functional Area Effects on Channel Type & Quantity

For the analysis of industry impacts on negotiation channels, 733 respondents revealed their industry affiliation. The impact of various industries is insignificant on the selection of the channel phone call (p-value .274, Phi & Cramer-V .151), face-to-face (p-value .339, Phi & Cramer-V .146), chat (p-value .680 Phi & Cramer-V .123), messaging (p-value .106, Phi & Cramer-V .169), audio (p-value .131, Phi & Cramer-V .165) and video conferencing (p-value .149, Phi & Cramer-V .163). However, the impact of various industries is significant on the selection and use of the negotiation channel email (p-value .011 Phi & Cramer-V .198). Specifically, the petrochemical and oil industries use more email in negotiations than statistically expected (standardized residual not quoted -2.1, quoted 1.5). Moreover, the automotive industry uses 15% less, and the pharmaceutical industry uses 23% less email than statistically expected. On average, negotiators across various industries use three negotiation channels (see Figure 13). Only negotiators in telecommunications or retail use almost four channels (3.6 mean value) while negotiating. Channel use in the remaining industries is scattered closely together between mean values of 3.4 to 3.1. Conversely, reduced channels are used in the public sector (2.7 mean value). Nonetheless, the differences in channel quantity among groups of industries are insignificant (Test of Variance Homogeneity p-value based on mean .862, ANOVA p-value between the groups .315).

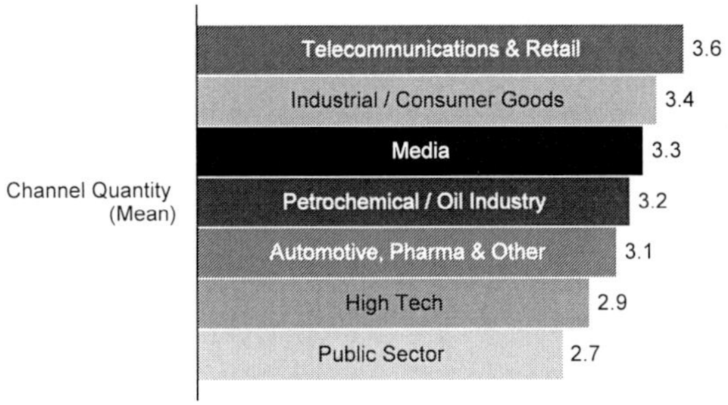

Figure 13. Channel Quantity Applied by Industry

The impact of the respondent's functional area of work on selecting the negotiation channels phone, face-to-face, chat, video conferencing, and other is insignificant. Partially significant is the impact of the functional area on the selection of messaging (Pearson Chi-Square p-value .066, Phi & Cramer V .142) and email (Pearson Chi-Square p-value .087, Phi & Cramer V .137) for negotiations. Despite this finding, the impact of the functional area on the selection of audio conferencing for negotiations is significant (Pearson Chi-Square p-value .044, Phi & Cramer V .147). On average, three negotiation channels are used by respondents from various functional areas in this study. Most negotiation channels (3.4 mean value) are used by business development teams, closely followed by purchasing and procurement (3.3 mean value), and the least media (2.6 mean value) is employed by the marketing department and project management (2.7 mean value). The category other (3.2 mean value) comprises the functional areas of communication, contract management, human resources, and information technology.

4.4.2 Effects of Negotiator's Position on Channel Type & Quantity

The relationship between a professional position of the negotiator and the following channel use is insignificant:

- phone call (p-value .103, Phi & Cramer-V .120),
- face-to-face (p-value .294, Phi & Cramer-V .100),
- email (p-value .244, Phi & Cramer-V .104),
- chat (p-value .608, Phi & Cramer-V .078),
- messaging (p-value .181, Phi & Cramer-V .110).

The link between the negotiator's professional position and the channel use of video conferencing during bargaining is partially significant (p-value .099, Phi & Cramer-V .121). However, the affiliation between the channel audio conferencing and the organizational position of the negotiator is significant (p-value .012, Phi & Cramer-V .150). Specifically, department heads use double the amount of audio conferencing in negotiations than expected (standardized residual not quoted -1.4, quoted 3.2). In comparison to other job profiles, department heads also utilize most channels (3.5 mean value). Whereas managing directors and divisional managers use equal negotiation channels (3.2 mean value). On average, team members and negotiators in the position of other use 3.1 negotiation channels. The least negotiation channels are deployed by team leaders (2.8 mean value), followed by managers without a team (2.9 mean value). For a pretest, the impact of the negotiator's position on the channel quantity, grounded in the Tests for Variance Homogeneity, is insignificant (p-value based on mean .226) resulting in data revealing homogenous variances. Nevertheless, differences between the positions are significant (ANOVA p-value .029). The influence of the negotiator's professional position on the channel quantity does not happen randomly with an ANOVA effect size of Eta-Square .019, which is minor in effect. In paired comparisons of positions with each other, no variances can be identified as values are insignificant.

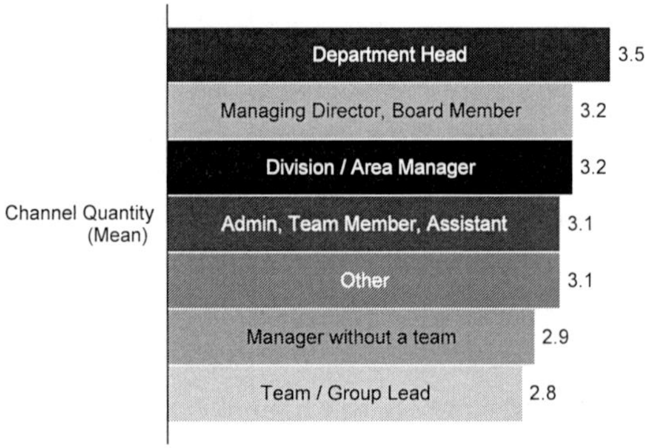

Figure 14. Channel Quantity Applied by Managerial Position

4.5 Preference for a Negotiation Channel

Ideally, in negotiating dyads, participants share their preference for a negotiation channel with the negotiation partner (Kachelmeier, 2002; Mennecke et al., 2000; Schoop, 2010). Within the scope of this study, the preference for a negotiation channel is analyzed. Results show that face-to-face (65%) is the most desired media for negotiators within this study (see Figure 15). In line with Eidhoff (2017), this result displays that technology alone is not adding value and that the individual is still an indispensable asset. A total of 709 cases were analyzed with the Monte Carlo simulation due to many small, expected values (Confidence Interval 99%, sampling quantity 10.000) to assess if there is a relationship between gender and its impact on the channel preference. Six cells (33.3%) display an expected frequency smaller than five, and the minimal expected frequency is .25; therefore, a precise significant test is needed. Values of Pearson Chi-Square only apply if not more than 20% of the cells display an expectancy value smaller than five, and the most minor expected frequency needs to be larger than one. The impact of gender on the channel preference is significant (Exact Test according to Fisher-Freeman-Halton p-value .014). In a detailed analysis, women prefer phone calls less for negotiations, whereas men prefer phone calls for negotiations more than expected. Men favor email in negotiations as opposed to women. Additionally, women desire significantly less face-to-face negotiations (standardized residual -2.0) than men. Female negotiators also select less video conferencing than their male counterparts. In terms of digital negotiations, women prefer to chat less, while men prefer to chat more. Messaging & audio conferencing are equally less preferred by females, whereas men use both channels slightly more than expected.

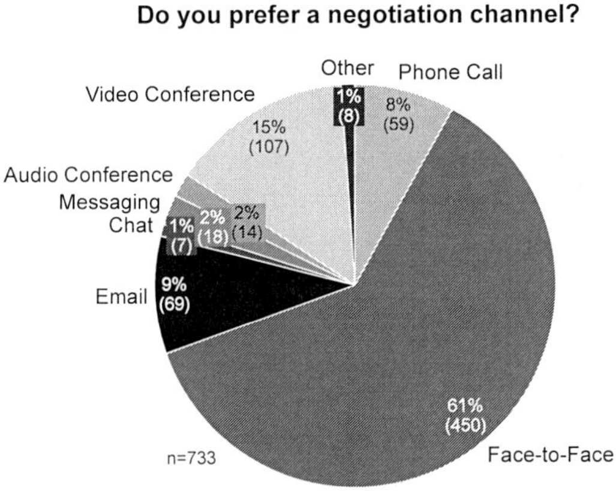

Figure 15. Survey Respondent's Preference for a Negotiation Channel

4.6 Factors Influencing Channel Choice

4.6.1 Gender Stimulus on Channel Choice

The impact of gender on the selection of seven negotiation channels was tested with a sample size consisting of 709 negotiators. It can be concluded that the effect of gender on the channel selection phone call (p-value .014) and face-to-face (p-value .001) is significant. According to Cohen (1988), the Phi-Coefficient and Cramer V show no effect between gender and the selection of the channel type phone call (Phi, Cramer's V .092). In contrast, the effect of gender on the selection of the channel face-to-face is weak (Phi, Cramer V .123). The impact of gender on the remaining five channels is insignificant. Men use the phone for bargaining more often than women. Women who quote (96 counts) to select the phone lies below the expected number (110 counts), whereas men use the phone (339 counts) more often than women in negotiations and even surpass the expected frequency of selecting this channel (325 counts). Men apply face-to-face more frequently in negotiations than women. The number of women who quote (98 counts) to select face-to-face lies below the expected score (116 counts). Men use face-to-face (362 counts) more often than women in negotiations and even surpass the expected number of this channel (344 counts). The corresponding p-value for the gender and face-to-face relationship is highly significant at .001. According to Cohen (1988), a small effect is identified between gender and the choice of the negotiation channel face-to-face (Phi, Cramer's V = .123).

4.6.2 Impact of Organizational Size on Channel Type & Quantity

A correlation between the organizational size and choice of channel was tested in a non-parametric Mann-Whitney-Test. The dependency between the size of the organization and three negotiation channels messaging, audio, and video conferencing is significant. Companies smaller in size (291.94 mean rank) use messaging (e.g., WhatsApp) for negotiations more often, which is a highly significant (<.001) correlation. Audio conferencing is more frequently quoted by larger companies (424.11 mean rank) and this relationship is also highly significant (<.001). Additionally, larger companies also quoted using video conferencing (376.44 mean rank), and this relationship is likewise significant at a p-value of .011. On average, companies of various sizes employ three negotiation channels (3.14 mean value) in their bargaining interactions. Interestingly, companies with 5.001 - 15.000 employees employ the lowest number of channels (2.89 mean value). Whereas companies with 15.000 - 50.000 employees use the highest number of channels (3.61 mean value). Smaller companies tend to use three channels on average, while medium-sized enterprises and large-scale corporations are trending towards using less than three channels. To this effect, it can be derived that the smaller the company, the more channels are employed, and the larger the company, the fewer channels are used. However, there is no correlation between the organizational size and the channel quantity (p-value .731, Spearman's rho Correlation Coefficient -.013).

4.6.3 Influence of Nationality on the Channel Selection & Quantity

The impact of nationality on the selection of negotiation channels was tested with an evaluation of 629 cases. Analysis reveals that the relationship between the nationality and the negotiation channel phone call (p-value .251), audio conferencing (p-value .116), and chat (p-value .088) are insignificant. On the contrary, the relationship between the nationality and the selection of face-to-face (p-value .018) or email (p-value .036) for negotiations is significant. In detail, North Americans use far less face-to-face (not quoted on 35 counts as opposed to the expected quantity of 20 counts) in negotiations than expected (standardized residual 3.2). This nationality sub sample quotes far less (23 counts) than expected (37.6 counts) the use of face-to-face in negotiations (standardized residual -2.4; Phi and Cramer's V = .178 weak effect). African negotiators use less email than expected (standardized residual 2.7) yet this relationship has a weak effect (Phi and Cramer's-V .169).

The relationship between nationality and messaging, or video conferencing is highly significant (p-value <.001). African (standardized residual 3.0) and South American negotiators (standardized residual 3.6) use more than double the amount of messaging in negotiations than expected. Northern Europeans decisively use less messaging (46) than expected (64.9 counts expected, standardized residual -2.4). The standard residual >2.0 demonstrates that all three nationalities are disproportionately represented in this distribution. This signifies that South Americans, and Africans more frequently and Western Europeans less frequently use selected messaging compared to an equal distribution. These three results are highly significant (p-value <.001) and through Phi and Cramer's-V this outcome is classified as a moderate effect (.280).

Moreover, African negotiators use less video than expected in negotiations (not quoted standardized residual 3.0, quoted standardized residual -2.3). On the contrary, Western European negotiators use disproportionately more video calls than expected (not quoted standardized residual -2.9, quoted standardized residual 2.2) and this effect is moderate (Phi and Cramer's-V .260).

On average, various nationalities in this study use 3.2 channels for their negotiations. Variance analysis identifies the highest number of negotiation channels used in South America (3.64 mean value), followed by Europe (3.25 mean value) and Africa (3.24 mean value), and the Middle East (3.17 mean value). Unpredictably, the least channels are used by North Americans (2.92 mean value) and Asians (3.01 mean value). The described deviations in channel quantity from the mean average are significant (p-value .012) according to the Test of Variance Homogeneity, which signifies a prevalence of variance distinctions. As a consequence of this outcome, the ANOVA testing procedure was replaced with the Welch Test for robust test procedures and identifies the equality of means is insignificant (Welch Test p-value .150) ending the test procedure.

4.6.4 Country of Operation Effects on the Channel Quantity

The impact of the country of operation on the selection of a phone call (p-value .121; Phi, Cramer's V = .125) and chat (p-value .317; Phi, Cramer's V = .104) for a negotiation is insignificant, while audio conferencing values are partially significant (p-value .080) with a minor effect (Phi and Cramer-V .132). A significant impact is caused by the country of operation on email negotiations (p-value .033; Phi, Cramer-V .145 small size effect according to Cohen (1988)). For this significant effect, the sub sample of African negotiators is responsible as they employ less email in their negotiations (14 counts) than expected (21 counts). In this case, the standardized residual is not normally distributed as it is at 2.4 for not quoting the bargaining channel email.

The link between the country of operation and video conferencing is significant (p-value .006; Phi & Cramer-V .167). The key driver behind this effect is that African negotiators use less video conferencing than expected (standardized residual quoted 2.4). Furthermore, the effect of the country of operation on the selection of the bargaining channel face-to-face is significant (p-value .002). This relationship is minor in effect (Phi & Cramer-V .181) according to Cohen (1988). For this result, the only significant effect is derived from negotiators in North America who use less face-to-face (30 actual counts) in negotiations than expected (47.1 expected counts; standardized residual not quoted 3.4; standardized residual quoted -2.5).

The coherence between the country of operation and messaging in negotiations is highly significant (p-value <.001). According to Cohen (1988), this relation displays an almost medium effect (Phi & Cramer-V .262). Negotiators in South America surpass the expectancy score (5.9 counts) for messaging in negotiations by 190% (16 counts). This is followed by African negotiators who exceed the expectancy score for messaging in negotiations by 115%. Due to the high standard residual larger than 2, the effects of negotiators in South America (standard residual not quoted -2.2 & quoted 4.2) and Africa (standard residual not quoted -1.5 & quoted 2.9) on messaging are, in isolation from each other, highly significant.

On average, three channels (3.18 mean value) are used in negotiations across various global regions. Reduced channels are used by negotiators operating in North America (2.91 mean value). In contrast, most channels are used by South American negotiators (3.7 mean value). Respondents conducting business negotiations in the Middle East use almost the same number of channels (2.97 mean value) as negotiators in North America. When negotiating in Asia (3.17 mean value) and Europe (3.19 mean value), almost identical number of channels are applied. The variations in mean for the number of negotiation channels employed across different global regions are insignificant (Tests for Variance Homogeneity p-value .242) making disparities in groups concerning channel employment insignificant (ANOVA p-value .114).

4.7 Structural Equation Modeling

4.7.1 The Model of Measurement

The evaluation of the model of measures classifies the reliability and validity of the constructs as satisfactory. Reliability assesses if the group of questions is interrelated by estimating measurement accuracy and reliability in an assemblage of items within the construct. The measurement of Cronbach's Alpha (α) serves as a quality criterion by testing if items represented in a construct hold internal consistency as an inter-item correlation. Although a factor loading above .7 is anticipated (Vinzi et al., 2010), scientists in social sciences regularly attain outer loadings beneath this threshold. Cronbach's Alpha for measurement of inner consistency in this model is over the threshold (.7) as an acceptable value, except in the self-efficacy (.665) and negotiation efficiency (.671) constructs. An improvement of Cronbach's Alpha by excluding singular items is only possible in the negotiation efficiency scale, yet a thorough investigation concluded not to eliminate any items on this scale for this study. The impacts of item deletion are carefully assessed to avoid mechanic omission of items (Latif et al., 2020). The omission of indicators with outer loadings from .4 to below .7 can modify not only composite reliability but also content and convergent validity. Item removal can lead to augmented composite reliability and average variance extracted (AVE) values above the anticipated threshold, which adds value to calculating the statistical construct (Hair et al., 2016).

According to Jöreskog (1971), Cronbach's Alpha is a conservative measurement (Trizano-Hermosilla & Alvarado, 2016), while Composite Reliability rho_c is a liberal measurement (Dijkstra, 2010 & 2014; Dijkstra & Henseler, 2015).

Composite Reliability rho_a is a compromise amid the measurement of Cronbach's Alpha and Composite Reliability rho_c. Appropriately, results indicate that items measure all latent constructs of the multi-channel negotiation model well. In this confirmatory modeling, both composite reliability rho_a and rho_c display elevated internal consistency amid reflective models with all values higher than .7. Construct indicators mostly exhibit values superior to the bottom threshold figure of .5 (Latif et al., 2020; Hair et al., 2010). The AVE reaches at least .5, except for the construct of the moderating effect complexity (.365) and the dependent variable construct of negotiation efficiency (.432). An AVE lower <.5 signals that less variance, and more errors are described by indicators in the construct (Al Majzoub, 2023). Nevertheless, a Convergent Validity with an AVE <.5 and a Composite Reliability (rho_c) >.6 is still adequate for a satisfactory Convergent Validity of this construct (Fornell & Larcker, 1981; Malhotra & Dash, 2011). In conclusion, most outer loadings are highly satisfactory (>=.7) and reach the reliability criteria of indicators.

Table 13. Construct Quality Criteria of Reliability and Validity

	Cronbach's Alpha	Composite Reliability (rho_a)	Composite Reliability (rho_c)	Average Variance Extracted (AVE)
Behavioral Dimensions				
Self-Efficacy	.665	.837	.808	.608
Agility	.772	.781	.854	.594
Synchronicity	.816	.818	.891	.732
Behavioral Control	.821	.823	.882	.650
Moderating Effects				
Complexity	.721	.726	.799	.365
Cognitive Resource Demands	.853	.941	.909	.770
Performance Risk	.809	.809	.887	.724
Dependent Variables				
Negotiation Effectiveness	.787	.819	.856	.552
Negotiation Efficiency	.671	.683	.790	.432

Discriminant Validity assesses the distinctiveness of constructs and examining this distinctiveness by exploring the correlation between these constructs is vital. In social sciences, constructs often overlap or are too highly correlated and for an examination if this is the case the Fornell-Larcker Criterion and the Heterotrait-Monotrait Ratio (HTMT) are discussed (Lobo, 2022). The Fornell-Larcker Criterion is a measurement for Discriminant Validity, which is rooted in correlation. In this model, indicators have a higher square root of the construct's AVE values on their own latent variables than all squared correlations with other constructs. This finding verifies that the manifest variables in each construct represent the latent variable and confirms its Discriminant Validity on the construct level. As an alternative to the AVE, the Cross-Loadings are analyzed to examine the Discriminant Validity of this reflective model. The Heterotrait-Monotrait Ratio values less than <.9 as a liberal figure or less than <.85 as a more conservative threshold establish the Discriminant Validity amongst this binary reflective model (Henseler et al., 2015; Hair et al., 2022). Essentially, all constructs are distinct and have their distinctiveness in this study. When assessing cross-loadings, all latent constructs load well on their parent factors. In consequence, other loadings with latent constructs are lower than with their parent factors. With loadings greater than 1 in this model, the distinctiveness of constructs is established. In conclusion, the quality criteria of the construct measures are achieved. The subsequent chapter presents the path analysis and significance of the structural model.

4.7.2 The Structural Model

The evaluation of the model structure analyzes the quality of its capacity to forecast endogenous constructs (Chau, 2018). For endogenous constructs, modifications in variance are visible in R-Squared values, which signifies "the model's explanatory power, also referred to as in-sample predictive power" (Hair et al., 2021; Rigdon, 2012; Sarstedt et al., 2021, p.23; Shmueli & Koppius, 2011). Simply put, it means how much dependent variable change is derived from independent variables (Hair et al., 2014). The appraisal of the model structure proves that behavioral dimensions contribute decisively (R^2 .290) to a more effective and moderate (R^2 .237) to a more efficient negotiation result (Cohen, 1988). In conclusion, the exogenous variables (R^2 >= .10) explain the variance of the construct adequately (Falk & Miller, 1992).

On the one hand, exogenous variables and moderating effects influence 29% of the variance in negotiation effectiveness. On the other hand, in negotiation efficiency, 24% of the variance is influenced by exogenous variables and moderating effects. Adjusted R-Squared values are slightly more conservative (for negotiation effectiveness R^2 .271 and negotiation efficiency R^2 .216) as these adapt to the number of terms in the model. Hence, adjusted R-Squared values only intensify if a new term advances the model fit above expectancy, which is not the case in this study. This study's adjusted R-Squared value decreases as the term does not improve the model fit enough.

Table 14. Coefficient of Determination for Effectiveness & Efficiency

	R-Squared	**R-Squared Adjusted**
Negotiation Effectiveness	.290	.271
Negotiation Efficiency	.237	.216

The typical characteristic of structural models is its reciprocal influences on variables. On that account, removing the exogenous variable(s) can affect the dependent variable(s). A removal of an exogenous variable from the model is labeled as F-Squared and equates into an R-Squared change. Opposingly, according to Cohen (1988), the removal of an exogenous variable from the construct does not improve the determination coefficient (R^2) as the effect size for all variables is F2 < .15, which is considered a small effect size.

The functionality PLSpredict appraises the model's prognostic competence (Hair et al., 2019, 2022). Henceforth, the Q^2 value in PLSpredict with the parameters number of folds 10, number of repetitions 10, random number generator fixed seed "compares the prediction errors of the PLS path model" against "simple mean predictions" (Channa et al., 2020, p. 30; Shmueli et al., 2016). To consider the forecast capacity of latent and manifest variables, various measurements of prediction error were undertaken, such as the root mean square error (RMSE), the mean absolute error (MAE), and the mean absolute percentage error (MAPE) (Shmueli et al., 2016).

Table 15. F-Squared for Independent and Moderating Variables

	Effectiveness	**Efficiency**
Self-Efficacy	.038	.017
Agility	.010	.030
Synchronicity	.002	.002
Behavioral Control	.020	.024
Complexity	.019	.014
Performance Risk	.003	.004
Cognitive Resource Demands	.009	.002
Cognitive Resource Demands x Agility	.001	.000
Performance Risk x Behavioral Control	.002	.003
Complexity x Synchronicity	.002	.000
Cognitive Resource Demands x Self-Efficacy	.001	.000
Performance Risk x Synchronicity	.000	.012
Complexity x Agility	.002	.000
Performance Risk x Agility	.001	.000
Complexity x Self-Efficacy	.002	.003
Cognitive Resource Demands x Behavioral Control	.000	.001
Complexity x Behavioral Control	.028	.018
Cognitive Resource Demands x Synchronicity	.008	.001
Performance Risk x Self-Efficacy	.003	.001

Table 16. Manifest Variable Prediction Summary

Manifest Variable	Q²predict	RMSE	MAE	LM_RMSE	LM_MAE
Effectiveness 01	0.145	0.653	0.441	0.663	0.460
Effectiveness 02	0.124	0.756	0.537	0.759	0.561
Effectiveness 03	0.124	0.748	0.553	0.755	0.568
Effectiveness 04	0.109	0.766	0.540	0.772	0.552
Effectiveness 05	0.044	0.985	0.823	0.986	0.813
Efficiency 01	0.009	1.009	0.851	1.022	0.864
Efficiency 02	0.076	0.909	0.705	0.906	0.715
Efficiency 03	0.083	0.805	0.566	0.809	0.594
Efficiency 04	0.044	1.029	0.861	1.026	0.864
Efficiency 05	0.072	0.974	0.796	0.969	0.776

The PLS model of behavioral dimensions in multi-channel negotiations has a superior forecast capacity as the positive Q^2 implies that the multi-channel construct's prediction error is minor compared to the prediction error of mean values (Mehrolia et al., 2023). According to Hair et al. (2013), the model has a moderate forecast relevance for both negotiation effectiveness (Q^2 .199) and negotiation efficiency (Q^2 .135). Overall, the model's forecast relevance, Q-Squared, is well reconstructed and proficient (>0).

Table 17. Latent Variable Prediction Summary

	Q²predict	RMSE	MAE
Negotiation Effectiveness	.199	.901	.638
Negotiation Efficiency	.135	.934	.701

4.7.3 Hypothesis Testing of Direct Relationships

Bootstrapping as a nonparametric procedure is used to test statistical significance. Through the Bootstrapping analysis (10.000 cases, bias-corrected and one-tailed) hypotheses of this study were tested. The parameter one-tailed is selected as the assumed relationship direction among the variables - positive or negative, strengthening or weakening - is known and grounded in the literature review. The weight of impact, also known as the beta coefficient, is expressed as the original sample and is significant if the beta coefficient is >.1. When using one-tailed bootstrapping, any t-value over 1.645 is significant.

Table 18. Direct Relationships of Behavior on the Negotiation Outcome

Causal Relationship	Original Sample	Sample Mean	STDV	T Statistics	P Values	
Self-Efficacy on Effectiveness	0.181	0.183	0.041	4.359	0.000	***
Self-Efficacy on Efficiency	0.125	0.125	0.039	3.200	0.001	**
Agility on Effectiveness	0.108	0.105	0.042	2.610	0.005	**
Agility on Efficiency	0.188	0.191	0.051	3.720	0.000	***
Synchronicity on Effectiveness	0.061	0.059	0.050	1.223	0.111	-
Synchronicity on Efficiency	-0.043	-0.044	0.052	0.829	0.203	-
Behavioral Control on Effectiveness	0.157	0.158	0.048	3.297	0.000	***
Behavioral Control on Efficiency	0.207	0.204	0.055	3.772	0.000	***

Note: ***p<.001, **p< .01, *p< .05

Collectively, 14 causal relationships are significant with a positive effect on the endogenous variables. The results reveal a positive impact of self-efficacy (beta value .183, t-value 4.359, p-value .000), agility (beta value .108, t-value 2.610, p-value .005), and behavioral control (beta value .157, t-value 3.297, p-value .000) on negotiation effectiveness. Apart from this, the effect of self-efficacy (beta value .125, t-value 3.200, p-value .001), agility (beta value .188, t-value 3.720, p-value .000), and behavioral control (beta value .207, t-value 3.772, p-value .000) on negotiation efficiency is also significant. Contrary to expectations, synchronicity does not positively impact negotiation effectiveness and efficiency. Henceforth, six of the initially formulated eight hypotheses related to direct relationships are confirmed.

Table 19. Confirmed Hypotheses of Direct Relationships on Outcome

#	Hypotheses	P-Value
H1	a) Self-efficacy positively impacts effectiveness. b) Self-efficacy positively impacts efficiency.	.000 .001
H2	a) Agility has a positive effect on effectiveness. b) Agility has a positive effect on efficiency.	.005 .000
H4	a) Behavioral control positively impacts effectiveness. b) Behavioral control positively impacts efficiency.	.000 .000

4.7.4 Comparing Direct Moderation Effects with Hypotheses

Investigating moderators clarifies potential indirect modifications in negotiation outcomes (Dardas & Ahmad, 2015). So far, research on moderator effects on the relationship between behavioral dimensions and the negotiation outcome in multi-channel negotiations is nonexistent. Moderation is about changing existing relationships (Ramayah et al., 2018) and by implication, changing an existing relationship requires more power. This study establishes and enhances amassed statistical power through the sample size. Along these lines, the high sample size (733 survey participants) acts favorably on this moderation analysis. Depending on the causal relationship, the moderators can strengthen and weaken the causal relationship's effects (Baron & Kenny, 1986). Testing of hypotheses reveals if moderating effects perform as a magnifier or buffer. Following a standard procedure, measures of complexity, cognitive resource demands, and performance risk "were centered in tests of moderation to maximize interpretability and minimize potential problems with multicollinearity" (Aiken & West, 1991; Connor-Smith & Compas, 2002, p. 47). As all constructs of independent variables and moderators are reflective and measured with a continuous scale, the product indicator approach is most suitable (Hair et al., 2017). Considering interaction terms in structural equations leads to measuring moderating variables with a continuous scale level (Hayduk, 1987; Jaccard & Wan, 1996). By assessing the stimulus of the variable interaction on the dependent variable, moderating effects are detected (Henseler & Chin, 2010). It encompasses "multiplying each indicator of the exogenous construct with each indicator of the moderator" (Chin et al., 1996; Ramayah et al., 2017, p. 7). If the result is meaningfully distinctive from zero, then moderation is proven. The path coefficient describes how the power of the independent on the dependent variable transforms with modifications in moderation (Eggert et al., 2005).

Moreover, it is necessary to consider modifications in R-Squared. In the primary effect model R-Squared amounts to .225 for effectiveness and .184 for efficiency (unadjusted). Considering all moderators in the interaction effect model R-Squared increases to .290 for effectiveness and .237 for efficiency (unadjusted). For effectiveness, adding the interaction terms results in an R-Squared increase of .065 – this means the additional variance amounts to 29%. For efficiency, the R-Squared increase is 29% (+.053). Accordingly, with moderators, the variance in each construct increases by approximately 30%.

Table 20. Additional Variance through Moderators

	Moderators Included	**Moderators Excluded**
R-Squared Effectiveness	.290	.225
R-Squared Efficiency	.237	.184

Among all moderators, complexity has the highest impact on variance, with an increase of .052 for effectiveness (5%) and .041 for efficiency (4%). Cognitive resource demands increase the variance for effectiveness by 3% (.029) and the variance for efficiency by .5% (.005). The impact of performance risk is minor, with a slight increase of .003 (.3%) on effectiveness and .012 (1.2%) for efficiency.

According to Kenny (2016), the effect size of complexity on effectiveness (.020) and efficiency (.010) indicates a medium effect. To such a degree, complexity has a medium effect on R-Squared change when removed as an exogenous variable (Cohen, 1988). Chin et al. (2003) declare that a F-Squared value with a subdued effect size "does not necessarily imply that the underlying moderator effect is negligible" (Ramayah et al., 2017, p. 18). A minor effect of interaction turns into meaningfulness if the corresponding changes in beta are consequential (Chin et al., 2003), whereas the direct effects of cognitive resource demands and performance risks on effectiveness and efficiency in the measure of F-Squared are negligible.

Bootstrapping reveals values for the interaction term with thresholds for highly significant (<=.01) at 2.33 and significant results (<=.05) at 1.645. Only complexity significantly impacts effectiveness and efficiency. Results for cognitive resource demands and performance risk are only partially significant. Complexity directly impacts the independent variable's effectiveness (beta value .142, t-value 3.175, p-value .001) and efficiency (beta value .127, t-value 2.778, p-value .003) positively. This effect is contrary to the original hypotheses, which postulated that complexity would have a negative effect on the causal relationships, whereas cognitive resource demands only positively impact effectiveness (beta value .087, t-value 2.313, p-value .10). Lastly, the effect of performance risk on efficiency is negative (beta value -.060, t-value 1.469, p-value .071). This converse, negative direction means the main effect is buffered.

Table 21. Direct Moderating Effects on Dependent Variables

Causal Relationship	Original Sample	Sample Mean	Standard Deviation	T Statistics	P Values	
Complexity on Effectiveness	0.138	0.143	0.044	3.138	0.001	**
Complexity on Efficiency	0.107	0.112	0.046	2.330	0.010	*
Cognitive Resource Demands on Effectiveness	0.086	0.085	0.037	2.309	0.010	*

Note: ***p<.001, **p< .01, *p< .05

4.7.5 Hypothesis Testing for Indirect Moderation Effects

Emanating from the theoretical work of Cohen (1988), complexity has a medium effect size (>.15) on the relationship between behavioral control and effectiveness (.032) and efficiency (.028). Contrastingly, Anguinis et al. (2005) argue “that the average effect size in tests of moderation is .009” (Hair et al., 2021, p. 162). Moreover, derived from Kenny (2018), an optimistic interpretation is that the effect size of complexity is sizeable on the relationship between behavioral control and effectiveness and of medium effect on efficiency. These F-Squared values confirm that the moderator contributes to the endogenous construct. The effect size of performance risk on the causal relationship of synchronicity and efficiency is with .015 of medium impact (Kenny, 2016). In contrast, cognitive resource demands only slightly affect the causal relationship of synchronicity and effectiveness (.008). Even this minor F-Squared effect size morphs into relevance if the conforming variations in beta are substantial (Chin et al., 2003).

Table 22. Indirect Moderating Effects on Dependent Variables

Causal Relationship	Original Sample	Sample Mean	STDEV	T Statistics	P Values	
Complexity on Self-Efficacy and Efficiency	-0.073	-0.070	0.042	1.749	0.040	*
Complexity on Behavioral Control and Effectiveness	0.186	0.185	0.042	4.420	0.000	***
Complexity on Behavioral Control and Efficiency	0.172	0.171	0.048	3.572	0.000	***
Cognitive Resource Demands on Synchronicity and Effectiveness	-0.099	-0.087	0.051	1.949	0.026	*
Performance Risk on Synchronicity and Efficiency	-0.131	-0.135	0.052	2.522	0.006	-

Note: ***p<.001, **p< .01, *p< .05

A moderator magnifies the relationship if the interaction term displays a positive beta (Aiken & West, 1991; Baron & Kenny, 1986; Warner, 2008). The analysis depicts that the moderating variable complexity acts as an amplifier on the relationship between behavioral control and effectiveness (beta value .182, t-value 4.137, p-value .000) and efficiency (beta value .152, t-value 3.359, p-value .000). This result is opposed to the original hypotheses which assumed that complexity would have a negative effect on negotiation effectiveness and efficiency. A substantial interaction term with a negative beta implies that cognitive resource demands buffer the relationship between synchronicity and effectiveness (beta value -.099, t-value 1.981, p-value .024). Also, the converse direction of performance risk on synchronicity and efficiency (beta value -.122, t-value 2.300, p-value .011) means that the main effect is reduced.

Three hypotheses regarding moderation effects in multi-channel negotiations can be confirmed as significant.

Table 23. Confirmed Hypotheses of Indirect Relationships

#	Hypotheses	P-Value
H5	b) The positive effect of self-efficacy on negotiation efficiency is weakened through complexity.	.040
H6	e) The positive effect of synchronicity on negotiation effectiveness is weakened through cognitive resource demands.	.026
H7	f) The positive effect of synchronicity on negotiation efficiency is weakened through performance risk.	.006

4.7.6 Controlling Variables in the Context of Channel Negotiations

Conventional statistical procedure requires consideration of controlling variables in causal structures to avert alternative reasons for significant statistical results or to minimize errors and intensify variable strength (De Battisti & Siletti, 2019). Modifications in a control variable do not occur. However, modifications in variable relationships and outcomes can be attributed to control. Understanding how to proceed with findings from control variable assessment is heterogenous (Breaugh, 2008; Spector et al., 2000). By analyzing control variables, the intention is to validate if variable interrelations are unadulterated (Becker, 2005; Spector & Brannick, 2011) to see if variable dependencies are inhibited by other variables (Breaugh, 2005, 2008; Meehl, 1970, 1971; Spector et al., 2000), and to successfully isolate an actual effect an exogenous variable has on the endogenous variable (Carlson & Wu, 2012, p. 415). Either an experiment can identify the impact of control variables through modification of sampling, or control variables of relevance are included in statistical analysis (De Battisti & Siletti, 2019). The second approach was applied in this study, where the specific effect of one control variable is statistically isolated from other variables (Becker, 2005). The PLS-based structural equation modeling analysis was run employing multi-group analysis once more with each of the subsequent controlling variables included in the model - gender, age, negotiation experience, negotiation history with the negotiation partner, industry, functional area, position, and organizational size. Hence, "comparing the results of this kind of analysis for hypothesized control variables allows a determination of the relative importance of different aspects of control" (Scholz & Schröner, 1999, p. 291).

Supplementary, it is necessary to consider modifications in R-Squared caused by control variables. In the main effect model with moderating variables R-Squared amounts to .290 for effectiveness and .229 for efficiency (unadjusted). Considering all moderators in the interaction effect model R-Squared increases to .318 for effectiveness and .270 for efficiency (unadjusted). For effectiveness, adding control variables results in an R-Squared increase of .028. This outcome signifies that the additional variance amounts to 2.8%. For efficiency, the R-Squared increase is 4.1% (.041).

Table 24. Comparison of R-Squared with Control Variables

R-Squared without Control Variables			**R-Squared with Control Variables**		
	R-Squared	**R-Squared adjusted**		**R-Squared**	**R-Squared adjusted**
Effectiveness	.290	.271	**Effectiveness**	.323	.274
Efficiency	.237	.216	**Efficiency**	.272	.218

For gender, age, and negotiation experience, an interaction effect was not identified. Minor interaction effects, partially significant (>.05), were uncovered for negotiation history, industry, position, and functional area. Major interaction effects, significant in result (<.05), were identified for the following control variables: organizational size, position, and industry. Henceforth, the greater the organization's size, the more advanced the effectiveness of the negotiation (p-value .009) making this correlation highly significant. Respondents acting as division head display fewer effective negotiations (-.292) than negotiators in other positions (p-value .009), followed by team leaders with less effective negotiations (-.215) than negotiators in other positions. Negotiators working in media or telecommunications show greater negotiation efficiency (p-value .016) than other industries. While respondents working with industrial goods show higher negotiation effectiveness (p-value .017), and negotiators from the public sector show an overall lower negotiation efficiency (p-value .035) than the industry standard. Group sizes strongly differed for some of the control variables, so the outcomes demand to be treated cautiously.

Table 25. Significant Control Variables

Control Variables	Original Sample	Sample Mean	Standard Deviation	T Statistics	P Values	
Organizational Size on Effectiveness	0.081	0.080	0.034	2.372	0.009	**
Divisional Head on Effectiveness	-0.292	-0.292	0.124	2.352	0.009	**
Media & Telecom on Efficiency	0.388	0.394	0.181	2.149	0.016	*
Industrial Goods on Effectiveness	0.265	0.256	0.125	2.127	0.017	*
Public Sector on Efficiency	-0.352	-0.342	0.194	1.810	0.035	*
Team Lead on Effectiveness	-0.215	-0.212	0.119	1.806	0.035	*

Note: ***p<.001, **p< .01, *p< .05

4.8 Cluster Analysis for Group Attributes

A cluster analysis is applied to identify groups with similar characteristics in the dataset (Hair et al., 2014). The dataset is divided into clusters based on the similarity of components in each subset (Everitt et al., 2001). This cluster analysis aims at grouping records to maximize the similarity of objects within clusters and minimizes the similarity between objects of different clusters (Milligan & Cooper, 1985).

During the scientific modeling of processes differentiating attributes are essential as criteria to assess if an observation with suspected causes relates or how to design the measuring sequence (Jaccard & Jacoby, 2010; Lee et al., 2012). It delivers answers to the question of how to measure criteria. A criterion is relevant to distinguish between different forms of multi-channel negotiations for decision-making. It defines the singularity and the affiliation to a certain cluster of multi-channel negotiations. Characteristics are determined, which define specific forms of the phenomenon. Herein, a criterion is a differentiating attribute as a prerequisite for a decision between two characteristic forms of a multi-channel negotiation. To define, investigate, assess, and compare the criteria of multi-channel negotiations, a catalog of criteria was compiled for a usability evaluation. Multi-channels meet criteria that are demanded within the communication process of the negotiation. As conducted by Danaher & Rossiter (2011), differentiating criteria are defined for the evaluation and choice of negotiation channels. By contextualizing media criteria to the context of multi-channel negotiations, the following criteria (see Table 26) are identified constructed around research of Gulati & Garino (2000); Harinck & De Dreu (2008); Prasarnphanich & Gillenson (2003a, 2003b), and Voeth & Herbst (2022).

Table 26. Criteria Definition for Categories of Multi-Channel Negotiations

Category	Criteria	Definition
Negotiation Organization	*Coordination*	The effort as valued consumption of resources to organize diverse single tasks or activities in complex multi-channel negotiations so that these harmoniously and meaningful as well as goal-oriented match each other.
	Alignment	An effort regarding used resources to align in the choice of channels.
	Documentation	The expense as an estimated expenditure of resources to document the interaction with diverse negotiation channels.
Negotiation Preparation	*Channel Strategy*	The channel strategy defines the type and quantity of utilized negotiation channels to attain the negotiation goal.
	Channel Tactic	To safeguard expediency and success of the negotiation, a regulated usage of negotiation channels is established.
	Channel Choice	The choice of one or multiple negotiation channels to execute a negotiation.
	Complementary	The coordination of negotiation channels for them to complement one another.
	Channel Quantity	The quantity of media channels in a negotiation.
	Process Orientation	The deployment of channels as a processual coordination.
Negotiation Conduct	*Steering*	The usage of negotiation channels enables an organizational and content-related steering of the negotiation.
	Adaptability	To lodge negotiation channels in changing negotiation circumstances to attain negotiation targets effectively and efficiently.
	Integrated Information Access	Integrated information access provides negotiators with the ability to access information available in one channel from another channel (adapted from Oh & Teo, 2006, p.35).
	Complexity	The more channels and diverse relationships between the channels, the more complex the multi-channel negotiation is.
	Fragmentation	The fragmentation of the negotiation into several negotiation episodes which are based on each other.

The table below accumulates the characteristics of differentiating criteria for the evaluation and choice of negotiation channels (Gulati & Garino, 2000; Harinck

& De Dreu, 2008; Prasarnphanich & Gillenson, 2003; Voeth & Herbst, 2022). The selected categorical criteria then lead to the classification of a group. The groups are divided into single-channel users or basic, intermediate, and advanced channel negotiations. This grouping approach provides a structured analysis and comparison of channel use in negotiations and corresponding behavior.

Table 27. Group Analysis of Negotiation Channel Quantities

Category	Criteria	Group1	Group 2	Group 3	Group 4	Group 5
		Single Channel	Basic Channels	Intermediate Channels	Advanced Channels	Expert Channels
Negotiation Organization	Coordination	low	medium	high	very high	highest
	Alignment	low	medium	high	very high	highest
	Documentation	low	medium	high	very high	highest
Negotiation Preparation	Channel Strategy	no	optional	optional	optional	optional
	Channel Tactic	no	optional	optional	optional	optional
	Channel Choice	low	medium	high	very high	highest
	Complementary	no	optional		optional	optional
	Channel Quantity	1	2	3	4	>=5
	Process Orientation	no	optional	optional	optional	optional
Negotiation Conduct	Steering	low	medium	high	very high	highest
	Adaptability	low	medium	high	optional	optional
	Integrated Information Access	no	optional		optional	optional
	Complexity	low	medium	high	optional	optional
	Fragmentation	low	medium	high	optional	optional

4.8.1 Multi-Group Analysis on the Channel Quantity Impact

The structural equation model was verified through a multi-group analysis, comparing different sample subgroups varying in size and composition. The defining differentiating criteria is the number of channels to form subgroups. Negotiation practitioners are keen to know the impact on a negotiation if the number of channels used augments. Through a multi-group analysis, the impact of channel quantities used during a negotiation on the outcome was examined to identify differences in results. This procedure enables the option to run the examination model for various groups and to examine the results for significant differences. The prerequisite for this analysis is identical measurement models (Huber et al., 2007, p. 51 and 118). The sample was divided into five subgroups of single, two-channel, three-channel, four-channel, and five and more channel negotiations. As a necessary condition that the relationships between constructs in the groups revert to the same construct content, the latent variables in all models are measured with the same indicators. With identical measurement instruments, an estimation of model parameters can be conducted (Huber et al., 2007) for each subgroup. It was then analyzed if the valuators in the subset present significant differences. The t-value was calculated for the difference between two path coefficients of differing samples.

Table 28. Estimation of Model Parameters with Five Subgroups

	Group 1	**Group 2**	**Group 3**	**Group 4**	**Group 5**
Type	Single Channel Negotiation	Basic Multi-Channel Negotiation	Intermediate Multi-Channel Negotiation	Advanced Multi-Channel Negotiation	Expert Multi-Channel Negotiation
Channel	1	2	3	4	5-8
N	73	188	198	160	114

4.8.2 Effectful Behavior in Dual-Channel Negotiations

With an increase in negotiation channels, generally, an increase in effects is documented up to the use of three channels. At its peak optimum of three negotiation channels, nine positive effects outweigh two adverse effects. From four negotiation channels with ten total effects, gradually these decline to only five total effects when using five to seven negotiation channels. For four negotiation channels, positive and negative effects are balanced. Even the use of five to seven channels identifies as more positive, with four positive effects and only one negative effect.

Behavioral dimensions are ineffective and unable to entirely influence the negotiation outcome if only a single channel is used. Hence, only one causal relationship is significant when testing single-channel negotiations in 73 cases during the multi-group analysis. Accordingly, a self-effective behavior in a single channel negotiation impacts effectiveness positively (beta value .329, t-value 2.044, p-value .020).

Table 29. Group 1: Negotiating with a Single Channel

Causal Relationship	Original Sample	Sample Mean	STDEV	T Statistics	P Values	
Self-Efficacy on Effectiveness	0.329	0.303	0.161	2.044	0.020	*

Note: ***p<.001, **p< .01, *p< .05

With an increase in channels, the impact of behavioral dimensions on the negotiation outcome increases. Effectiveness increases overall in importance, and efficiency only plays a role when using two channels. When using two negotiation channels, the impact of agility on negotiation effectiveness (beta value .311, t-value 2.958, p-value .002) and efficiency (beta value .173, t-value 1.800, p-value .036) is substantial. A controlled behavior acts favorably on negotiation efficiency (beta value .268, t-value 2.571, p-value.005). The prevalence of complexity leads to a more efficient negotiation (beta value .179, t-value 2.154, p-value .016). It strengthens the effect of behavioral control on efficiency (beta value .167, t-value 2.110, p-value .017) and effectiveness (beta value .159, t-value 1.897, p-value .029). If the negotiator fears that the multi-channel negotiation is a risk to the overall performance, this impacts the effect of synchronicity on efficiency negatively (beta value -.217, t-value 1.915, p-value .028). The fear of a lack in performance can harm the negotiation outcome.

Table 30. Group 2: Negotiating with Two Channels

Causal Relationship	Original Sample	Sample Mean	Standard Deviation	T Statistics	P Values	
Agility on Effectiveness	0.311	0.279	0.105	2.958	0.002	**
Behavioral Control on Efficiency	0.268	0.248	0.104	2.571	0.005	**
Complexity on Efficiency	0.179	0.185	0.083	2.154	0.016	*
Complexity on Behavioral Control and Efficiency	0.167	0.163	0.079	2.110	0.017	*
Performance Risk on Synchronicity and Efficiency	-0.217	-0.181	0.113	1.915	0.028	*
Complexity on Behavioral Control and Effectiveness	0.159	0.152	0.084	1.897	0.029	*
Agility on Efficiency	0.173	0.192	0.096	1.800	0.036	*

Note: ***p<.001, **p< .01, *p< .05

4.8.3 Behavioral Optimum in Three-Channel Negotiations

The importance of behavioral dimensions in steering the multi-channel negotiation augments with increasing negotiation channels to the count of three channels. In sum, 198 negotiators used three channels in their most recent, significant negotiation. In three-channel negotiations, five effects are positive and two negative on negotiation effectiveness. Three negotiation channels also act favorably on negotiation efficiency with four positive effects. Consequently, making the usage of three channels the optimum number of channels. A self-effective behavior shapes effective (beta value .269, t-value 4.255, p-value .000) and efficient (beta value .221, t-value 3.054, p-value .001) negotiations decisively. Moreover, agility drives efficient negotiation conduct (beta value .279, t-value 2.340, p-value .010). The importance of complexity on effectiveness (beta value .266, t-value 3.798, p-value .000) and efficiency (beta value .234, t-value 2.193, p-value .014) is confirmed. Complexity weakens the weight of synchronicity on the effectiveness of the negotiation (beta value -.243, t-value 3.093, p-value .001), while it strengthens the influence of behavioral control on negotiation effectiveness (beta value .174, t-value 1.925, p-value .027). An anticipated performance risk buffers an effective negotiation behavior (beta value -.120, t-value 1.795, p-value .036). On the contrary, it also strengthens the effect of agility on effectiveness (beta value .271, t-value 3.073, p-value .001) and self-efficacy on efficiency (beta value .133, t-value 1.887, p-value .030). The more cognitive resources are in demand in this three-channel negotiation, the higher the positive force on effectiveness (beta value .149, t-value 2.038, p-value .021).

Table 31. Group 3: Negotiating with Three Channels

Causal Relationship	Original Sample	Sample Mean	Standard Deviation	T Statistics	P Values	
Complexity on Effectiveness	0.266	0.285	0.070	3.798	0.000	***
Self-Efficacy on Effectiveness	0.269	0.265	0.063	4.255	0.000	***
Performance Risk on Agility and Effectiveness	0.271	0.266	0.088	3.073	0.001	**
Complexity on Synchronicity and Effectiveness	-0.243	-0.221	0.079	3.093	0.001	**
Self-Efficacy on Efficiency	0.221	0.222	0.072	3.054	0.001	**
Agility on Efficiency	0.279	0.271	0.119	2.340	0.010	*
Complexity on Efficiency	0.234	0.244	0.107	2.193	0.014	*
Cognitive Resource Demands on Effectiveness	0.149	0.154	0.073	2.038	0.021	*
Complexity on Behavioral Control and Effectiveness	0.174	0.171	0.090	1.925	0.027	*
Performance Risk on Self-Efficacy and Efficiency	0.133	0.130	0.071	1.887	0.030	*
Performance Risk on Effectiveness	-0.120	-0.134	0.067	1.795	0.036	*

Note: ***p<.001, **p< .01, *p< .05

4.8.4 Decline of Positive Effects in Advanced Channel Negotiations

Negotiations with four negotiation channels reveal five strengthening and, at the same time, five weakening effects. On that account, the advantages, and disadvantages of using this specific quantity of channels are balanced. A shift occurs from the former predominant influence on effectiveness to a strong impact on efficiency. Essentially, the number of substantial impacts on efficiency increases when using four channels. However, efficiency is more negatively influenced by variables using four as opposed to only using three channels. The impact on effectiveness decreases when using four negotiation channels compared to three with two positive and one negative impact. The behavioral impact on negotiation efficiency in this channel setting is less favorable with three positive and four negative variable relationships. In comparison to using a lesser number of channels, the positive impact of behavioral dimensions in steering a multichannel negotiation decreased when four negotiation channels are used in 160 cases. Both self-efficacy on efficiency (beta value .234, t-value 2.496, p-value .006) and on effectiveness (beta value .184, t-value 1.885, p-value .030) and behavioral control for efficiency (beta value .252, t-value 2.199, p-value .014) and for effectiveness (beta value .326, t-value 2.673, p-value .004) are concrete pillars in maneuvering the negotiation into safe waters with high efficiency and effectiveness values.

In five instances, all weakening effects are caused by moderators. With four channels, the relevance of performance risk increases. Performance risk deteriorates the efficiency of the negotiation (beta value -.157, t-value .1.746, p-value .040). It reduces the positive weight of self-efficacy on the effectiveness of the negotiation (beta value -.180, t-value 1.861, p-value .031) and subsides the effect of synchronicity on efficiency (beta value -.217, t-value 1.944, p-value .026). On the one hand, cognitive resource demands strengthen agile behavior, which favorably impacts efficiency in the negotiation (beta value .285, t-value 1-760, p-value .039); on the other hand, cognitive resource demands weaken the ability of a synchronized negotiation behavior with conveyance and convergence processes and deteriorate efficiency (beta value -.267, t-value 1.861, p-value .031). Overall, complexity as an influencer becomes less critical as opposed to other channel settings. In comparison, in this four-channel constellation complexity reduces an agile behavior in the negotiation and diminishes efficiency (beta value -.201, t-value 1.864, p-value .031).

Table 32. Group 4: Negotiating with Four Channels

Causal Relationship	Original Sample	Sample Mean	Standard Deviation	T Statistics	P Values	
Behavioral Control on Effectiveness	0.326	0.357	0.122	2.673	0.004	**
Self-Efficacy on Efficiency	0.234	0.231	0.094	2.496	0.006	**
Behavioral Control on Efficiency	0.252	0.270	0.115	2.199	0.014	*
Performance Risk on Synchronicity and Efficiency	-0.217	-0.237	0.112	1.944	0.026	*
Self-Efficacy on Effectiveness	0.184	0.204	0.098	1.885	0.030	*
Cognitive Resource Demands on Synchronicity and Efficiency	-0.267	-0.245	0.143	1.861	0.031	*
Complexity on Agility and Efficiency	-0.201	-0.171	0.108	1.864	0.031	*
Performance Risk on Self-Efficacy and Effectiveness	-0.180	-0.159	0.097	1.861	0.031	*
Cognitive Resource Demands on Agility and Efficiency	0.285	0.272	0.162	1.760	0.039	*
Performance Risk on Efficiency	-0.157	-0.133	0.090	1.746	0.040	*

Note: ***$p<.001$, **$p< .01$, *$p< .05$

4.8.5 Behavioral Control – Decisive for Expert Channel Negotiations

Variable impacts are more geared towards efficiency with two positive and one negative effect than effectiveness with two positive effects in negotiations from five and more channels.

Table 33. Group 5: Negotiating with 5-7 Channels

Causal Relationship	Original Sample	Sample Mean	Standard Deviation	T Statistics	P Values	
Behavioral Control on Efficiency	0.492	0.466	0.141	3.489	0.000	*
Behavioral Control on Effectiveness	0.364	0.406	0.136	2.673	0.004	**
Self-Efficacy on Effectiveness	0.277	0.328	0.145	1.917	0.028	*
Agility on Efficiency	0.283	0.263	0.150	1.886	0.030	*
Cognitive Resource Demands on Self Efficacy and Efficiency	-0.312	-0.195	0.175	1.779	0.038	*

Note: ***p<.001, **p< .01, *p< .05

Evidently, more control is required of negotiators (beta .492, t-value 3.489, p-value .000) when using five to seven channels in 114 cases, positively impacting the overall efficiency and effectiveness of the negotiation. Additionally, a self-effective behavior is vital for effectiveness (beta .277, t-value 1.917, p-value .028). Using five to seven channels visualizes that cognitive resource demands increase for negotiators. With this, a self-effective behavior lessens, leading to an overall efficiency loss (beta -.312, t-value 1.779, p-value .038).

4.8.6 Selective Behavior Strengthens Effectiveness & Efficiency

The most substantial direct effect on effectiveness is achieved by complexity (p-value .000) and self-efficacy (p-value .000) using three negotiation channels. Agility proves to be an essential driver of effectiveness when using two channels (p-value .002). Throughout channel usage, self-efficacy continues to provide a stable and positive impact on effectiveness if negotiators are dealing with single (p-value .020) or four (p-value .030) and more channel negotiations (p-value

.028). The same applies to behavioral control, which displays a solid impact on effectiveness (p-value .004) for four or more negotiation channels.

Performance risk directly reduces effectiveness (p-value .036), while cognitive resource demands augment it (p-value .021) in three-channel negotiations. Indirectly, the highest impact on effectiveness, is triggered by performance risk and complexity. Using three channels, performance risk augments agility effects on effectiveness (p-value .001), yet complexity reduces synchronicity effects on effectiveness with a p-value of .001. Both in single (p-value .029) and triple channel negotiations (p-value .027), complexity drives behavioral control on effectiveness, while performance risk buffers self-efficacy on effectiveness (p-value .031) in four-channel negotiations.

The highest direct drivers on negotiation efficiency are behavioral control for dual- (p-value .005), quadruple- (p-value .014), and quintuple-channel negotiations (p-value .000), as well as self-efficacy for triple- (p-value .001) and quadruple-channel negotiations (.006). Throughout the use of three to five and more channels, agility yields favorable effects in three (p-value .010), four (p-value .036), and more channels (p-value .030) on efficiency. Complexity is a stable driver for efficiency using two (p-values .016) to three (p-value .014) channels. However, when applying five or more channels, performance risk is a direct buffer to efficiency (p-value .040). Complexity has the highest indirect impact on efficiency. Using two channels, complexity acts favorably on behavioral control and efficiency (p-value .017); nonetheless it minimizes agility and efficiency (p-value .031) when using double the number of channels. Throughout dual (p-value .028) to quadruple channel usage (p-value .026), performance risk diminishes synchronicity effects on efficiency. Surprisingly, with three channels, performance risk acts favorably on self-efficacy and efficiency (p-value .030). On the one hand, cognitive resource demands reduce synchronicity (p-value .031) and self-efficacy effects (p-value .038) on efficiency in 4-5 channel negotiations. On the other hand, cognitive resource demands improve agile effects on efficiency (p-value .039) using quadruple channels. With an increase in negotiation channels, the importance of behavioral dimensions in steering multi-channel negotiations increases. As a result, the optimum behavioral steering capacity is reached at three channels. Effective steering of the multi-channel negotiation with the selected behavioral dimensions decreases when using four or more negotiation channels. In opposition, only the importance of behavioral control increases with a growing number of channels.

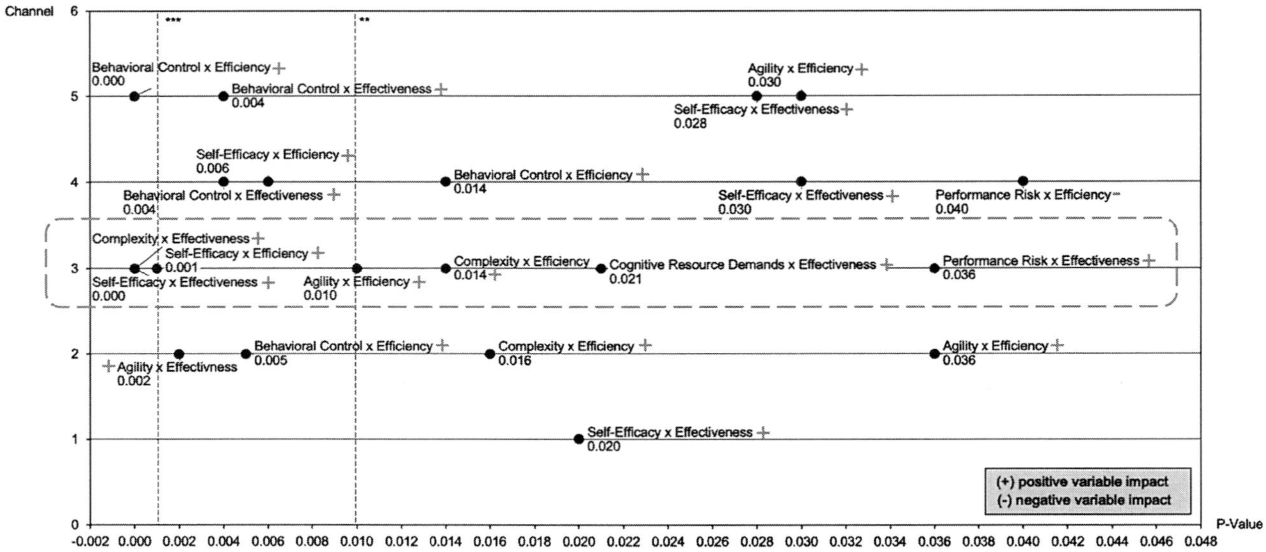

Figure 16. Direct Behavioral Effects Across Channel Quantities

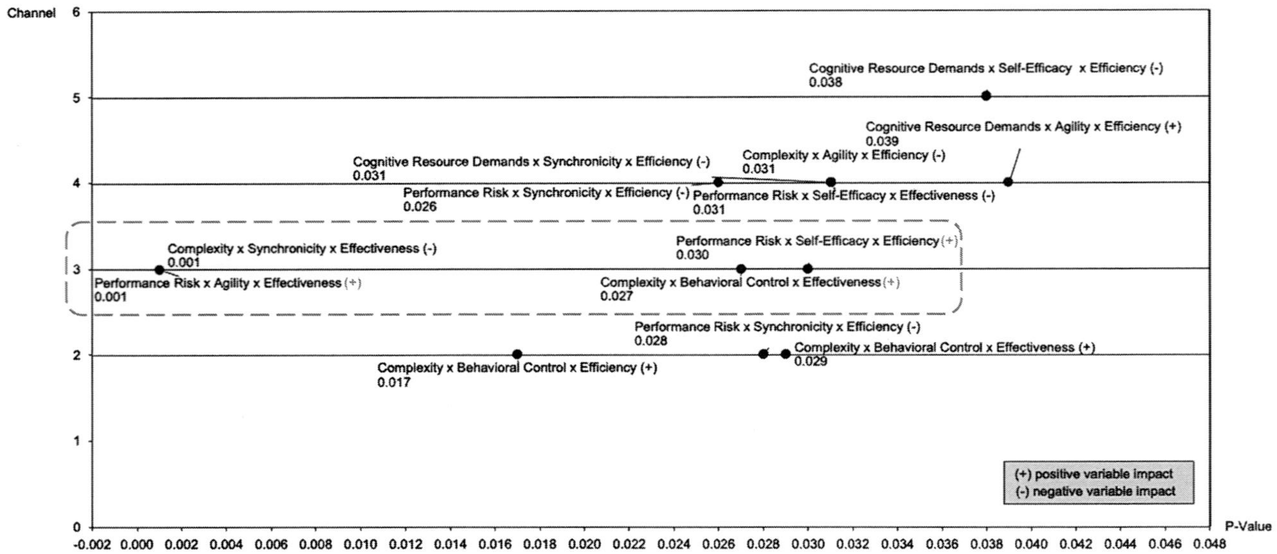

Figure 17. Indirect Moderating Effects on Behavior and Outcome

5 Discussion

This study presents global applications of multi-channel negotiations in business practice. Initially, trends of using specific channel types and quantities in multi-channel negotiations are discussed and evaluated. Accordingly, individual drivers for intentional channel selection are reviewed. Supplementary, the use of negotiation channels along the negotiation stages is presented and examined. Notably, the channel preferences of global negotiators are surveyed. In this context, behavioral dimensions' influence on the effectiveness and efficiency of multi-channel negotiations are assessed, and strategic implications are drawn from it. This study is extensive in its reach as it processes input on multi-channel negotiations from 733 international negotiators in diverse global business markets. The accumulated results of this study have substantial implications on negotiation science and applied negotiation management.

5.1 Strategic Process Steering through Channel Selection

Interdisciplinary research denotes the terminology multi-channel as a sequenced or parallel use of multiple communication media to send a message effectively and efficiently to a target group (Barley, 1991; Greenhalgh & Chapman, 1995). Transferring this concept of multi-channel communication and marketing (Cummins et al., 2016; Mallen, 1973; Manser Payne et al., 2017; Park & Lee, 2017; Verhoef et al., 2015) to negotiation translates into a modification of the negotiation process through modes of communication media (Bazerman & Moore, 2013; Putnam & Carcasson, 1997; Putnam & Roloff, 1992). In a multi-channel negotiation, more than two communication channels – analog, digital, or both – are used during various stages. Necessarily, to reach a consensus in this environment of complex negotiations, managers employ multiple negotiation channels synergistically in line with the principles of Integrated Marketing Communication Theory (Kitchen & Schultz, 1999). Several scholars emphasize the importance of comprehending the reasons behind media selection (Maity & Dass, 2014; Pookulangara et al., 2011; Wang et al., 2015; Yu et al., 2011). This study reveals that multi-channel combinations do not happen randomly, as 67% of study participants select a negotiation channel intentionally. This result aligns with theoretical recommendations for the targeted use of negotiation media for success (Adair & Loewenstein, 2013; De Dreu & Van Kleef, 2004a, 2004b; Dwyer, 2018; Salacuse, 2005). Subsequently, half of the respondents in this study influence the negotiation process outcome by selecting media channels. Adding to that, only a few negotiators (11%) believe they can impact the economic outcome of a negotiation with a channel. Even less negotiators (9%) focus on shaping the socio-emotional outcome of the negotiation with a specific channel choice. Opposed to the general assumption that "channel characteristics largely determine channel choice," the media selection in this study is determined by the process-, economic and socio-emotional outcome (Macik et al., 2012; Verhoef et al., 2007; Wang et al., 2015, p. 390). For negotiators, the reasons for actively influencing the negotiation process are versatile. Depending on the negotiation, the most appropriate channel mix is utilized (Crump, 2011; Crump & Druckman, 2012) with three fundamental reasons for channel selection to steer the process outcome reported by study participants: 1) negotiation context, 2) managing the negotiation, or 3) managing behavior. Of the three categories, negotiation management, as a trigger for channel selection, is the most critical to global negotiators. Mayer & Voeth (2022) confirm constructive effects of a systematic management of the negotiation. Successively, 200 respondents named negotiation management an austere cause for consciously choosing negotiation channels. This implies the significance for a strategic negotiation management (Mayer & Voeth, 2022; Voeth & Herbst, 2015; Voeth & Mayer, 2018). Essential drivers to manage the negotiation are, for instance, implementing and following through on a negotiation strategy and corresponding tactics as part of the negotiation preparation (Voeth & Herbst, 2015). Furthermore, documenting the negotiation progression is essential to managing the bargaining session. In line with Voeth et al. (2020), active control over time management is a pressing need for negotiators, and lastly, convenience is an aspect that negotiators focus on when proactively managing the negotiation through channel selection. Subsequently, 92 respondents identified the negotiation context as a principal reason to select negotiation channels intentionally. This finding supports the work of Ambrose et al. (2008) that channel choice is partially influenced by contextual factors. Finally, 93 respondents named behavior as a critical objective in selecting negotiation channels on purpose to influence the process outcome. Following these insights, limiting the choice of media is counterproductive to the negotiation process and outcome.

5.2 Unpuzzling the Rationale behind the Channel Quantity

Multi-channel negotiations substitute single-channel negotiations completely, as 96% of negotiators use at least two negotiation channels throughout their recent significant negotiation. This conclusion conforms with findings of Dennis et al. (2008), Shahriza & Heckman (2005), and Watson-Manheim & Bélanger (2007) that solitary communication media is inferior in effectiveness as opposed to multi-channel. Negotiators even prefer dynamic and challenging negotiations (Putnam & Holmer, 1992), with favorable two to four negotiation channels applied by 75% of survey respondents. Men use slightly more than three negotiation channels, whereas female negotiators prefer fewer than three. Reasons behind limited channel use among females could be the exposure of non-cooperative behavior through multiple media (Carnevale et al., 1981; Lewis & Fry, 1977). Channels may spotlight dominance or exploitation, which might reinforce competition among negotiators. It could also be those women, contrary to men, follow a more controlled negotiation style with fewer channels.

A significant correlation is that the older the negotiator, the more channels are used. This phenomenon could be linked to the experience of a senior negotiator and specifically according to Schwarz (2012) experience in dealing with (channel) information. On the contrary, younger negotiators might settle negotiations faster with fewer channels. Negotiation experience is vital for an individual cognitive negotiation knowledge management (Bazerman et al, 1985; Thompson, 1990). If negotiation experience is prevalent, then negotiators tend to use approximately one additional channel (3.1 mean value) as opposed to a lack of experience (2.2 mean value). The more experience a negotiator has, the more channels are operated.

Differences in the application of channel quantities between professional positions of negotiators are significant. The most channels, a total of four, are employed by negotiating department heads. The position of a department head often requires active negotiation through diverse channels. Specifically, department heads use double the amount of audio conferencing in negotiations than expected. These intense audio negotiations could be induced by the familiarity with the negotiation partner, where audio functions are sufficient to close negotiations, or it could be triggered by an active avoidance of visual access in line with the Barrier Effect by Lewis & Fry (1977) and Carnevale et al. (1981).

Managing directors and divisional managers use an equal amount of negotiation channels (3.2 mean value). In both high-ranking positions, the amount and intensity of negotiation tasks require a minimum of three channels to close the deal. Control variable analysis reveals that, through an interaction effect, respondents acting as division heads display fewer effective negotiations than negotiators in other professional positions. This result could be produced by a possible hierarchical induced distance to the negotiation context as division heads act on a higher strategic level as opposed to experts involved in implementing negotiating operations. Team leaders deploy fewer than three negotiation channels, followed by managers without a team. This reduced channel usage could be due to fewer negotiating tasks on this management level. Furthermore, control variable analysis reveals that, through an interaction effect, team leaders also negotiate less effectively than negotiators in other positions. This issue could be due to a strategic distance paired with an enhanced operational focus, which could limit negotiation performance.

Three is the privileged and most trusted channel number for successful global negotiations. Global negotiators use, on average, three media channels. South American negotiators implement up to four negotiation channels, the most used among global negotiators. European, African, Middle Eastern, and Asian negotiators use three solid channels to seal the deal. Contrary to initial assumptions, less than three negotiation channels are used by North Americans. Reasons for either a high or low number of negotiation channels could be determined by local legislation, requirements of the work environment, or even cultural aspects.

5.3 The Driving Forces for Negotiation Media Selection

5.3.1 Classic Negotiation Media Retain Dominance

External factors and the level of digitization in business transform negotiations, and with this, new forms and channels of negotiation emerge (Lax & Sebenius, 1986; Weiss, 2006; Zartman, 1994, 2003). Although digital negotiations are prominent and well-established, it is needful to assess if traditional negotiation channels "retain their historically favored attributes of trust and reliability" (Danaher & Rossiter, 2011, p. 6). However, the results validate that new negotiation channels do not replace classical negotiation media. Digital communication channels in negotiation range from social media to collaboration tools such as Microsoft Teams. This finding contradicts the assumption that "with the rapid and widespread uptake of new electronic media channels, traditional communication channels are expected to decline" (Danaher & Rossiter, 2011, p. 7). In effect, chat and messaging still play a minor role in negotiation.

After decades of negotiation research, the most prominent unknown territory is the effects of a mixture of various media in a negotiation (Ambrose et al., 2008; Damian et al., 2008; Dennis et al., 2008). The majority of negotiators in this research use the identical three negotiation channels as opposed to Agndal's et al. (2017) assumption that the choice of negotiation media today is vast compared to decades ago. As a result, the top three selected negotiation channels are email, face-to-face, and video conferencing. This finding matches the concept of Synchronicity Theory in which, depending on the negotiating process, either synchronous or asynchronous media is selected to achieve a state of conveyance (information sharing) or convergence (consensus building) (Lind & Zmud, 1991). In each case, the channel affects the clarity of the information exchange (Sitkin et al., 1992). For information sharing approvingly, asynchronous email is selected, while in consensus building, synchronous face-to-face or video conferencing is favored.

As digital media are the global game changers (Danaher & Rossiter, 2009), the negotiation stages greeting & initiation, background & information exchange, agenda setting, and the final agreement formulation (Yuan et al., 2003) are predominantly conducted via email. In line with the Barrier Effect, email provides fewer distortion effects as opposed to visual or audible media (Carnevale et al., 1981; Carnevale & Isen, 1986; Lewis & Fry, 1977; Pruitt, 1981). Similarly, email acts as a barrier "to reduce interpersonal conflict and competition" (Stuhlmacher & Citera, 2005, p. 71). During these specific negotiation stages, "face-to-face interactions are minimized to improve group process and performance" (Dalkey, 1969; Stuhlmacher & Citera, 2005, p. 71; Delbecq & Van de Ven, 1971). Due to travel and cost advantages, countless negotiations settle electronically (Chiu et al., 2005; MacDonald & Smith, 2004; Poole & DeSanctis, 1990; Shakun, 2005). Opposed to the study of Danaher & Rossiter (2011), email is not rated negatively by respondents of this study. Findings confirm that email is more trustworthy. Also, email is convenient, acceptable, and appropriate. Hence, email is quick and efficient; with this effect, involved negotiators are more concentrated on the negotiation content. Overall, negotiators tend to be more accommodating toward numerous email messages.

Conversely, at a singular negotiation stage, the issue discussion is primarily conducted via face-to-face or video conferencing. This outcome is in line with the Media Richness Theory by Daft & Lengel (1986), in which the output of an assignment progresses when informative demands correspond with the richness of a channel. Face-to-face and video conferencing can send rich information in ambiguous settings, and email navigates through low informative assignments (Dennis et al., 2008). Furthermore, it validates studies of McGrath (1993) that face-to-face is the ideal medium for negotiating conflicts of interest, followed by video conferencing with a marginal fit as the information transfer tends to be leaner. Negotiators prefer to see their counterparts. Survey respondents repeatedly identified that the reason for this is to react in time to any nonverbal communication appropriately. With all senses present in a face-to-face setting, it is easier to steer the negotiation after picking up nonverbal cues strategically on to safer grounds. This finding supports the Social Presence Theory (Short et al., 1976) and the Psychological Distance Perspective (Citera et al., 2005; Stuhlmacher & Citera, 2005) compiled by Geiger (2020) that "other communication media than face-to-face lack in social presence" (p. 217). By choosing face-to-face, negotiators avoid "reduced social cues" (Geiger, 2020, p. 217; Kiesler & Sproull, 1992; Sproull & Kiesler, 1986) through psychological distance (Dorado et al., 2002; Kiesler et al., 1984; Morris et al., 2002; Stuhlmacher & Citera, 2005; Wellens, 1986, 1989) and actively seek social impact (Latané, 1981). Overall, these findings strongly confirm the conclusion of Damian et al. (2008) that a combination of lean and rich media improves negotiation effectiveness during negotiation stages.

Previous research demonstrates negotiator's distinct preferences for channels (Flanagin & Metzger, 2001; Rice et al., 1998; Westmyer et al., 1998; Yuan et al., 2003) in accordance with the Theory on Uses and Gratification by Katz et al. (1973). When survey participants were specifically asked about the most desired negotiation channel, results illustrated that the classic negotiation medium face-to-face (65%) is still the front-runner. Again, this is undoubtedly triggered by the requirement to see the negotiation partner. Respondents strongly argued that necessary nonverbal behavior could be observed by visually capturing the opponent. This visual observation aids in decisively navigating the negotiation process successfully.

Highly significant differences occur in the importance of media channels for information sharing. Accordingly, the most salient communication channels for sharing information are face-to-face, video, and email. Nevertheless, video and audio conferencing, email, and phone calls are equally well perceived by global negotiators for information sharing. As expected, messaging and chat are the least valuable information-sharing channels. For the objective of consensus building during the negotiation, highly significant differences in assessing the importance of media channels occur. For consensus-building, the most important channels are face-to-face, video, and phone calls. In line with the research of Baltes et al. (2002), the negotiator's choice for decision-making is preferably conducted face-to-face. The main difference between the purpose of information sharing and consensus building is that instead of email, the respondents opt for a phone call to arrive at a mutual decision. Subsequently, it can be confirmed that all three channels display audio functions which are more intimate and therefore, inherently crucial for decision-making. Opposed to the information-sharing results, the decision-making channels are not interchangeable. In line with channel results for information sharing, the channels chat and messaging are the least useful for consensus building.

5.3.2 Gender, Age and Experience Stimulus on Media Selection

Survey data reveals that men more regularly use the phone for negotiations than women. In line with the Barrier Effect, audio-only communication reduces "attacking behavior in a competitive negotiation" (Carnevale et al., 1981; Carnevale & Isen, 1986; Lewis & Fry, 1977; Pruitt, 1981; Stuhlmacher & Citera, 2005, p. 71). This channel preference could be caused by a confrontative negotiation style requiring a direct, real-time audio exchange. Also, male negotiators could rely more on voice-only negotiations than female negotiators. Hypothetically, female negotiators could fear a loss of steering capacity and corresponding control when using phone calls in negotiations.

Moreover, male negotiators habitually interact more face-to-face than female negotiators. As discovered by Carnevale et al. (1981), Carnevale & Isen (1986), Lewis & Fry (1977), and Pruitt (1981), this could be due to a reduction of hostility when visual access is removed. Motivations to avoid face-to-face bargaining are a drop in dominance and competition, leading to a win-win atmosphere (Pruitt, 1981). A confrontative negotiation style could also trigger men to use face-to-face more often than females. Finally, the negotiator needs to be more active in observing the other party visually to steer the deal into a closing. Captivatingly, the impact of gender on channel preference is significant. Opposed to statistic expectations, women prefer all negotiation channels - phone, email, face-to-face, video conferencing, chat, messaging, and audio - less than statistically expected. In contrast, men startlingly prefer all negotiation channels more than statistically expected. This finding could be due to a general avoidance of women using multiple channels. Possibly, men could be more inclined towards applying various channels.

Senior negotiators apply significantly more video conferencing than the younger generation. This result may occur because the older negotiators feel in control when visually receiving the other end. However, it could also be that younger negotiators feel content with using nonvisual media to close the deal. Additionally, more experienced negotiators use video conferencing, phone calls, and email more regularly than negotiators with experience to a lesser extent. The triad of video, phone, and email proves successful; subsequently, experienced negotiators use it more habitually than less experienced negotiators.

5.3.3 Negotiation Media Induced by Industry & Organizational Size

Specifically, the petrochemical and oil industries use more email in negotiations than expected. As bargaining is an exchange between dyads or more participants with partially different preferences over one or more objects (Lewicki et al., 2015; Voeth et al., 2015), this phenomenon could be due to the negotiation object, which can be well negotiated via this specific asynchronous media as opposed to media use in other industries. Opposingly, less email than expected is employed by the automotive (-15%) and pharmaceutical (-23%) industries. This finding could hint at complex negotiation objects requiring more sophisticated media to conclude the negotiation. Through control variable analysis, an interaction effect reveals that respondents working in media or telecommunications display greater negotiation efficiency than other industries. Respondents working with industrial goods perform better in negotiation effectiveness, and negotiators in the public sector show an overall lower negotiation efficiency than the industry standard.

Companies smaller in size cite the use of messaging for negotiations more often. This result could be induced by a less conventional, more innovative, and agile negotiation mode. Furthermore, smaller companies are often in sole proprietorship, which equates to less regulation overall. With fewer employees, fewer locations, and lesser meetings, the choice could be to select the fastest way to negotiate via convenient messaging. Small companies characterized by unbureaucratic lesser regulating processes do not discourage negotiation messaging. The use of audio and video conferencing is preferred more by larger companies. This circumstance may be traced back to larger companies usually using collaboration tools such as Microsoft Teams or online conferencing, such as Zoom or Skype, which enable video or audio conferencing from any location. Large corporations' communication and negotiation processes shift into communication platforms. Control variable analysis identifies an interaction effect for the respective company size and negotiation effectiveness. As a result, the more expansive the organization's size, the more advanced the overall effectiveness of the negotiation. This movement powerfully translates into the fact that negotiators in larger companies are more effective than smaller enterprises.

5.3.4 The Geographic Divide in Employment of Negotiation Channels

Generally, an exploration of the various nationality impacts on channel choice demonstrates that North Americans and Western Europeans negotiate commonly with classical negotiation media, specifically email and video conferencing. This dynamic could be triggered by various legislation and company regulation towards media use. At the same time, South Americans and Africans have a strong tendency to negotiate with new negotiating channels. This direction results in South Americans and Africans more frequently and Western Europeans less frequently selecting messaging for negotiations compared to an equal statistical distribution. Again, this result could be triggered by a more tolerant legislative environment in these specific geographic regions or even novel approaches toward channel use in negotiations. Respectively, negotiators in South America even surpass the expectancy target for messaging in negotiations by 190%. African negotiators exceed the probability target for messaging in negotiations by 115%. The effects of South American and African nationalities on messaging, seen in isolation from each other, are highly significant.

Furthermore, African negotiators use less video and email than expected in negotiations, whereas in contrast, negotiators in Western Europe use disproportionately more video calls than expected. As no single communication technology will likely prove to be adequate for all task situations or phases of a single task (Mennecke, 2000), video conferencing is easily and rapidly configurable by users to support multiple communication modalities. Desktop video conferencing is a technology that blends voice, video, and synchronous data sharing. In such an environment, users can be empowered to choose the most appropriate media configuration for a given situation, for instance, formal or informal communication (Fish et al., 1993). Also, this technology allows users to move between lean and rich communication easily and rapidly during the same communication episode (Alavi et al., 1995).

Moreover, disproportionately more video calls can also stem from the pandemic in which corporate travel bans restricted face-to-face negotiations. Also, the home office dynamic still prevails and strongly encourages convenient video negotiations. Surprisingly, North Americans use far fewer face-to-face negotiations than expected. This progression could also be a post-pandemic result in which video conferencing continues to be exponentially promoted. Additionally, ongoing corporate cost-cutting initiatives undoubtedly encourage the permanent shift from physical confrontations to digital exchanges.

5.4 Behavioral Dimensions

The established model of behavioral dimensions and moderating effects in this research exhibit solutions to Lasswells' Communication Theory (1948) appeal to decipher the sender and receiver of an impactful exchange through media – and contributes decisively (29%) to a more effective and moderate (24%) to a more efficient negotiation result. The model displays a well-reconstructed and proficient strength in its ability to forecast negotiation effectiveness and efficiency results. In line with the Integrating Marketing Theory by Kitchen & Schultz (1999), the effectiveness and efficiency of a behavioral negotiation strategy on its outcome are measured using data and analysis to improve and optimize the strategy continually. The results reveal a positive influence of self-efficacy, agility, and behavioral control on negotiation effectiveness. Apart from this, the effect of self-efficacy, agility, and behavioral control on negotiation efficiency is also significant. This discovery fulfills the demand of managers, as Diamond & Weiss (2013), Lewicki & Litterer (1985), Raiffa et al. (2002), Shell (2006), and Watkins (2014) rightly outline, for specified measures to improve negotiation results. Contrary to expectations, synchronicity does not positively impact negotiation effectiveness and efficiency. This dynamic proves that behavioral dimensions are a lever that directly impacts the outcome of a negotiation. Hence, behavior is decisive for the success of the negotiation.

An interaction term is used to test the structural equations for moderation (Hayduk, 1987), as all relevant moderators in this study have a continuous scale (Jaccard & Wan, 1996). Overall, moderating effects only slightly enhance the variable impact on efficiency by 5.3% and effectiveness by 6.5%. Among all moderators, complexity is the key driver and has the highest impact on variance for effectiveness (+5%) and efficiency (+4%) with a medium effect size. Conversely, performance risk and cognitive resource demands are minor in effect. Bootstrapping reveals that only complexity significantly impacts effectiveness and efficiency positively. Complexity acts as an indirect amplifier of the association concerning behavioral control and effectiveness and has a medium influence on the bond between behavioral control and efficiency. This complexity weakens the effect of self-efficacy on efficiency. Performance risk indirectly buffers synchronicity's effect on efficiency, confirming the analogous hypothesis, while cognitive resource demands indirectly reduce the effect of synchronicity on effectiveness. This result also ratifies the formulated initial hypothesis.

Substantially, communication performance improves as negotiators apply an assortment of channels to conduct bargaining assignments instead of singular channel employment. Regardless of the positive effects, behavioral dimensions are ineffective and unable to entirely influence the negotiation outcome if a solitary channel is used. With an increase in negotiation channels, the importance of behavioral dimensions in steering the multi-channel negotiation augments. Schwarz (2012) identifies a deficit in planning and steering negotiations leading to a loss of effectiveness and efficiency potentials. Subsequently, when using more than one channel, behavioral impacts on effectiveness increase, and suddenly, behavioral impacts on efficiency play a relevant role. When using two negotiation channels, agile negotiators, within the scope of this study, perform strongly on both negotiation effectiveness and efficiency. Further, controlled behavior enhances efficiency in the negotiation. This new two-channel complexity leads to higher efficiency levels overall and strengthens behavioral control's effect on efficiency and effectiveness. Fear of the negotiator towards possible multi-channel risks leads to a collective weakening of synchronicity on efficiency.

5.4.1 Performance Optimum in the Negotiation Channel Triad

The optimum behavioral steering capacity is reached at three channels. The positive effects outweigh the adverse at its peak optimum of three negotiation channels. A self-effective behavior shapes effective and efficient negotiations decisively toward a positive outcome. Supplementary, enhanced agility is a driver for efficient conduct in the triad. Transferring work of Daft & Lewin (1993), Marschak & Reichelstein (1998), and Tushman & Nadler (1978) to multi-channel negotiations translates into efficiency of an agile negotiator by grasping uncertainty and changes through reacting with an adaptive bargaining style. Outstandingly, complexity acts as a driver for better negotiation effectiveness and efficiency within this channel trinity.

Nevertheless, complexity also deteriorates the effect of synchronicity on the effectiveness of the negotiation. In contrast, it strengthens the weight of behavioral control on negotiation effectiveness. Unpredictably, an anticipated performance risk acts favorably on effective behavior in negotiations. Performance risk also strengthens the effect of agility on effectiveness and self-efficacy on efficiency. In three-channel negotiations, more cognitive resources are in demand, which in sum, translates into a positive impact on effectiveness. The channel triad allows for the cognitive encoding of the negotiation message. Hence, the level of thinking and attention a negotiator devotes to a three-channel negotiation acts positively on effectiveness.

5.4.2 Performance Decline with more than four Negotiation Channels

In four-channel negotiations, positive and negative effects are balanced. Compared to using fewer channels, the positive impact of behavioral dimensions in steering decreases when four negotiation channels are used. A shift occurs from the former positive effects on effectiveness to efficiency. However, efficiency is more negatively influenced. Both self-efficacy and behavioral control are concrete pillars in maneuvering the negotiation into success with high efficiency and effectiveness values. With four channels, the relevance of performance risk increases. Accordingly, performance risk generally weakens negotiation efficiency and diminishes self-efficacy on effectiveness and the effect of synchronicity on efficiency. On the one hand, cognitive resource demands strengthen agile behavior, which favorably impacts efficiency in the negotiation. On the other hand, cognitive resource demands weaken the ability of a synchronized negotiation with conveyance and convergence processes and weaken overall efficiency. Captivatingly, complexity as an influencer overall becomes less critical. Nevertheless, complexity still weakens agile negotiation behavior, which diminishes efficiency vastly.

Effective steering of the multi-channel negotiation with the selected behavioral dimensions decreases when using more than four negotiation channels. However, only the importance of behavioral control increases with a growing number of channels. Nevertheless, even the use of five or more channels identifies as more favorable as opposed to the disadvantages. More control is demanded of negotiators when using five or more negotiation channels, which positively impacts the overall efficiency and effectiveness of the negotiation. Additionally, self-effective behavior proves vital for success in negotiation effectiveness. With increased cognitive demands in five or more channel negotiations, the self-effective behavior declines, leading to an overall efficiency loss.

5.4.3 Prime Impetus for Negotiation Effectiveness & Efficiency

An in-depth analysis of effectiveness drivers shows that the most potent direct effect is achieved by complexity and self-efficacy using three negotiation channels. Additionally, agility is a decisive positive force on effectiveness, especially when using only two channels. The agile negotiator transforms challenges into prospects (Alavi et al., 2014; Pitafi et al., 2018; Sherehiy, 2008). This result aligns with Kumar & Motwani (1995) that agility accelerates activities on critical paths and is a direct indicator of time-based effectiveness (Voeth et al., 2020). Furthermore, throughout various channel quantity usage, self-efficacy continues to provide a stable and positive impact on effectiveness regardless of whether negotiators deal with single or multi-channel negotiations. The identical aspect applies to behavioral control, which displays a stable influence on effectiveness yet more on advanced multi-channel negotiations from four or more channels. Performance risk directly reduces effectiveness, while cognitive resource demands augment it in three-channel negotiations. Indirectly the highest impact on effectiveness is triggered by performance risk and complexity. Using three channels, performance risk augments agility effects on effectiveness; in contrast, complexity reduces synchronicity effects on effectiveness. Both in single and triple-channel negotiations, complexity drives behavioral control on effectiveness, while performance risk buffers self-efficacy on effectiveness in four-channel negotiations.

In summary, the highest direct drivers on negotiation efficiency are behavioral control for two to five-channel negotiations and self-efficacy for three to four-channel negotiations. Throughout the use of three and more channels, agility yields favorable effects on efficiency through proactive resilience toward uncertainty and change. Likewise, complexity is a stable driver for efficiency using two to three channels. In contrast, performance risk is a direct buffer to efficiency when applying five or more channels. In accordance, complexity has the highest indirect impact on efficiency. Using two channels induces complexity to act favorably on behavioral control and efficiency yet minimizes agility and efficiency when using double the number of channels. Throughout dual to quadruple channel usage, performance risk diminishes synchronicity effects on efficiency. Surprisingly, with three channels, performance risk acts favorably on self-efficacy and efficiency. On the one hand, cognitive resource demands reduce synchronicity and self-efficacy effects on efficiency in four- to five-channel negotiations; on the other hand, cognitive resource demands improve agile effects on efficiency using four channels. In conclusion, with an increase in negotiation channels, the importance of behavioral dimensions in steering multi-channel negotiations increases. Notably, the optimum behavioral steering capacity is reached at three channels. This finding fulfills the request of companies to retrieve measures to enhance their negotiation performance (Fisher et al., 2011; Lewicki & Hiam, 2001; Mnookin et al., 2017; Shell, 2006; Thompson, 2011). Effective steering of the multi-channel negotiation with the selected behavioral dimensions decreases when using four or more negotiation channels. Only the importance of behavioral control increases with a growing number of channels. Conspicuously, there is no positive impact of synchronicity, nor an indirect enhancement through synchronicity on the outcome variables, thus making it irrelevant in impact. Conversely, complexity proves surprisingly positive in its impact on the multi-channel negotiation outcome.

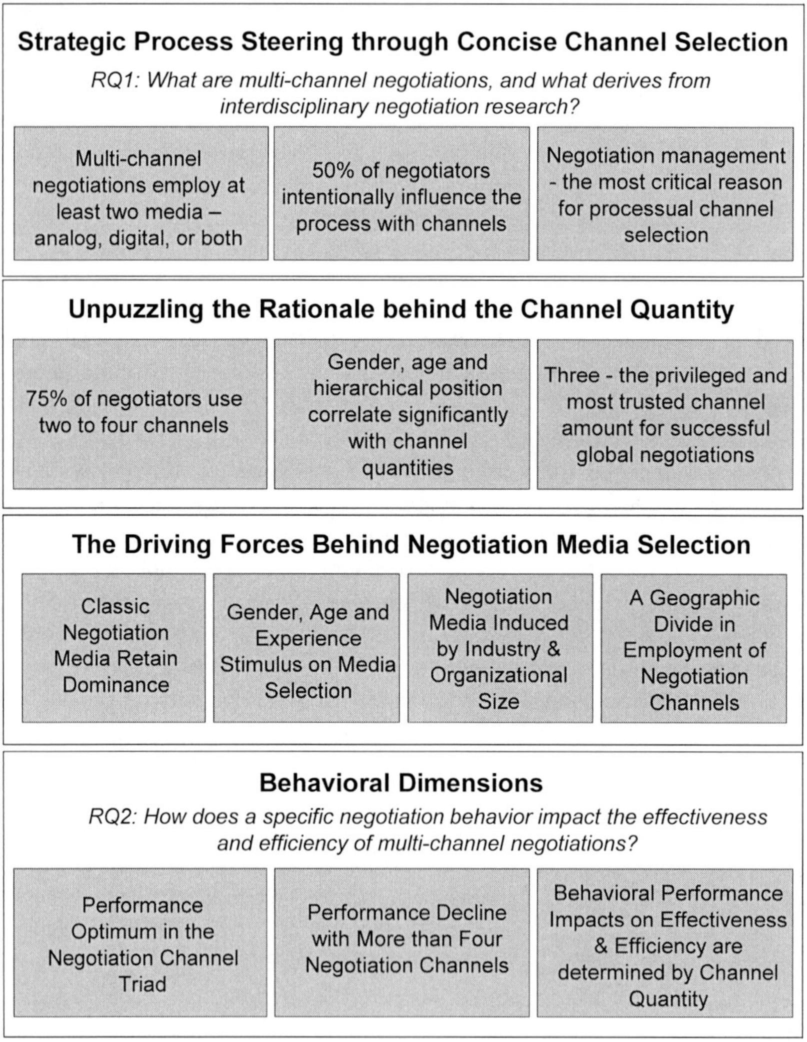

Figure 18. Overview of Main Results

6 Conclusion

The existence and need for multi-channel negotiations are identified in a systematic literature review over a publication period of 30 years. A multi-channel negotiation consists of a minimum of two negotiation channels – analog, digital or both – used during the negotiation phases. Relevant strategic communication theories on impactful interactions and efficiency assessed within the scope of this study apply to a broad communication spectrum, and not only to negotiations. Generally, insights regarding negotiation media indicate that impacts depend on the context, and no blanket statement is possible. Largely, analysis of the effects of negotiation channels on the process and outcome of multi-channel negotiations reveals contradictory findings. A partial explanation for these contradictory findings is the analysis of negotiation media over three decades in which communication media evolved making it problematic to derive a successful recipe for multi-channel negotiations from these. Accordingly, transferring general findings from communication research to a specific negotiation process requires validation to advance negotiation research further. This research identifies an effect mechanism of behavioral dimensions - self-efficacy, agility, synchronicity, and behavioral control - in multi-channel negotiations. The study tests the perceived effectiveness and efficiency of negotiation messages delivered by this specific negotiation behavior through different channels. For this purpose, negotiation managers specifically evaluated the effectiveness and efficiency of negotiation behavior in multi-channel negotiations. To broaden the research view, this study tests whether a certain quantity of channels is more ideal for achieving optimal negotiation results. In accordance, the observed effectiveness and efficiency of behavior is contrasted with the respective channel quantity for analysis of positive effects on the negotiation outcome.

Analysis of survey data reveals the strategic direction of channel selection and channel quantities that are deployed. On the one hand, this creates transparency on media preferences; on the other hand, it produces strategic motives for intentional media use. For a holistic comprehension, an overview presents negotiation media choices during the various negotiation phases. A causal analysis quantifies the effects of behavioral dimensions and moderators on outcome effectiveness and efficiency in multi-channel negotiations through a structural equation model. The measurement of behavior and resulting performance aid in classifying a particular behavior for successful negotiations (Stoshikj, 2014). This approach validates the individual negotiator's perceived effectiveness and efficiency of a negotiation funneled by a specific behavior through different channels. A certain number of channels is ideal for achieving optimal negotiation results. Overall, the following results shed light on the strategic employment of multi-channel negotiations.

6.1 Formation of Empirical Theory on Multi-Channel Negotiations

This research study establishes a substantial positive empirical theory by bringing theoretical and observed variables into a relationship with corresponding rules (Bagozzi, 1998). It conceptually examines multi-channel negotiations, merges fundamentals across different disciplines to create a solid theoretical base and derives conclusions for negotiations. Established on thoughts of Purdy et al. (2000), Sheffield (1995), and Swaab et al. (2012), this study applies communication media as the context for negotiation behavior. It examines key media channels comprehensively in correspondence with behavioral dimensions of negotiators on the economic criteria of relative effectiveness and efficiency of a negotiation (Heinen, 1976; Hyder & Prietula, 2000; Katz & Kahn, 1978a, 1978b, 1978c; Macharzina & Wolf, 2008; Neely et al., 2005; Purdy et al., 2000; Schmidt, 2020).

Specifically, communication behavior impacts the success of a negotiation. Since the 1990s, negotiation behavior has replaced economic profit as the most prominent variable in negotiation research (Carnevale et al., 1981; Carnevale & Isen, 1986; Geiger, 2020; Stuhlmacher & Citera, 2005). Subsequently, negotiation behavior is analyzed, as an independent variable, with its effect on the defined performance measures negotiation effectiveness and efficiency, as dependent variables, in multi-channel negotiations. In line with Albers & Eggert (1988), the content of the specified behavior is theoretically deducted by dividing the respective constructs into the following dimensions: self-efficacy, agility, synchronicity, and behavioral control. Selected moderating effects specific to a multi-channel negotiation in the framework of this study are complexity, cognitive resource demands, and performance risk. A conceptual framework defines the effect mechanism of these behavioral dimensions and interrelated moderating effects in multi-channel negotiations. From this conceptual model, reciprocal hypotheses have been derived. As academics urge to back theory on communication behavior with large quantities of generated data, a quantitative study yields results on general media use and the behavioral effect mechanism in multi-channel negotiations. This study uses data from 733 international negotiation experts in companies who assess the possible success factors in multi-channel negotiations. With this, empirical evidence validates the conceptual model to demonstrate triggers for modification in negotiation performance.

This exploration reinforces the argument that classical interpretations of logical channel selection and assignment conduct do not fully explain task success (Mennecke, 2000). Results suggest that issues beyond normative characteristics of tasks and technology influence users' communication preference and performance. Accordingly, it was found that the specified behavioral dimensions affect up to 30% of negotiation effectiveness and 24% of efficiency. Additionally, the model incorporates requirements of real-life complexity in multi-channel negotiations (Braun & Hadwich, 2016). On the one hand, within the context of behavioral dimensions, complexity directly impacts effectiveness and efficiency positively. On the other hand, it acts as an indirect strengthening amplifier on behavioral control and effectiveness as well as on the relationship between behavioral control and efficiency. However, in line with the prior hypothesis, complexity weakens self-efficacy effects on efficiency.

6.2 Management of Multi-Channel Efficiency & Effectiveness

McLuhan (1973) declares that the medium is the message, which means a channel expansion has different effects on the human sensorium. With augmented channels, the importance of behavior in steering the multi-channel negotiation intensifies. The strongest direct behavioral drivers on negotiation efficiency in general are self-efficacy and behavioral control. Additionally, enhanced agility is also a continuous driver for efficient conduct in negotiations. Contrary to general belief, complexity is not only a stable direct driver for efficiency yet also acts favorably on behavioral control and efficiency. However, it minimizes agility and efficiency when using double the number of channels. Negotiation science implies measurable positive effects of coaching techniques for successful negotiations (Movius, 2008; Nadler et al., 2003). Notably, the study results are relevant for an expansion on negotiation training (Bazerman et al., 1985; ElShenawy, 2010; Herbst et al., 2019; Voeth et al., 2015) to enhance a self-effective, agile, and controlled negotiation behavior in multi-channel negotiations.

Opposed to the initial hypothesis, a performance risk acts favorably on self-efficacy and efficiency in a channel triad. The risk of failure triggers a stronger belief in one's ability. In contrast to triad channel results, when applying five or more channels the fear of risking performance output is suddenly a direct buffer to efficiency. In this case, the negotiator believes there is a potential level of risk that the multi-channel negotiation does not yield the expected results. Moreover, throughout the channel usage a performance risk diminishes synchronicity effects on efficiency. This implies that perceived risk does not allow synchronicity to emerge and unfold positive effects on negotiation efficiency. In this manner, performance risk suppresses the emergence of a coordinated synchronous behavior with a shared goal to create a win-win situation or even to close an integrative negotiation. An environment with minimized channels must prevail for synchronicity to take effect. Likewise, cognitive resource demands reduce synchronicity and self-efficacy effects on efficiency in advanced channel negotiations yet also improve agile effects on efficiency using four channels. In conclusion, performance risk and cognitive resource demands are both distortion effects to synchronicity in its positive impact on the negotiation.

Throughout media usage, self-efficacy continues to provide a stable and positive impact on effectiveness regardless of negotiators dealing with a single or four and more channels. Self-efficacy is the belief that one's effort equates into a projected outcome (Volkema et al., 2013). In this sense, self-efficacy determines adjustments in behavior and, from this, results in behavioral performance (Bandura, 1982). This result supports Kirk and Brown (2003) that in this conceptual model the independent variable self-efficacy explains the most variance in negotiation performance.

6.3 The Optimum Performance in Three-Channel Negotiations

The optimum behavioral steering capacity is reached at three channels. This finding validates theory that coordinating the same media content constantly from diverse channels has more impact (Wang, 2006). Especially a self-effective behavior, with an overall most potent effect, shapes effective and efficient negotiations decisively towards a positive outcome. Uncommonly, complexity proves positive in its impact on the negotiation outcome and also acts as a strong driver for better negotiation effectiveness and efficiency. Effective steering of the negotiation with the selected behavioral dimensions decreases as the optimum of three channels is exceeded. Nevertheless, the positive effect of behavioral control increases with growing channels. A successful management of channels and complex patterns of interrelationships can be achieved by augmenting behavioral control. While using four channels, self-efficacy and behavioral control are substantial pillars in maneuvering the negotiation into success with high efficiency and effectiveness values. As Bandura (1977) pointed out, with self-efficacy the individual is willing to invest resources and perseverance toward the end achievement. Behavioral control instead displays a stable effect on four or more channels and positively augments effectiveness. More control is demanded of negotiators when using five or more negotiation channels, which positively impacts the overall efficiency and effectiveness of the negotiation. Nevertheless, with increased cognitive demands in five or more channel negotiations, the self-effective behavior lessens, thus leading to an overall loss in efficiency. Throughout the use of three to five and more channels, agility yields favorable effects on efficiency. This adopted flexibility, responsiveness, adaptation, quickness, and reconfiguration deals successfully with highly complex, turbulent, and uncertain environments (Amos, 1998; Dove, 2001; Herrmann et al., 2006; Kidd, 1994; Sharifi et al., 2001; Sharifi & Zhang, 1999). In conclusion, this leverage effect of multiple negotiation channels produces negotiation agility. Conspicuously, there is no positive impact of synchronicity, nor an indirect enhancement of synchronicity on the outcome variables, making it irrelevant in impact. Synchronicity needs space to thrive on establishing conveyance by constructing a mental model to form joint comprehension. Under certain conditions, negotiations do not allow these subprocesses to thrive. Even applying different negotiation channels in each subprocess does not support conveyance and convergence. Opposed to theory of Dennis et al. (2008), a shared pattern of coordinated behavior never emerges.

In line with Walther (1992), negotiator behavior transcends media, which is opposed to McLuhan's theory. Success in a negotiation is attributed, among other aspects, to the negotiator behavior. Divergent from several authors (Delaney et al., 1997; Staples & Zhao, 2006; Zornoza et al., 2002), this study identifies differences in behavior while using media. Delaney et al. (1997), Geiger (2020), Griffith & Northcraft (1994), Yang (2012), and Zornoza et al. (2002) detect, for negotiations specifically, no disparity in behavior using channels. This scientific research is valuable to business negotiations, as it identifies multi-channel effects on negotiation behavior, which until today have rarely been part of research, thus implying the need for further investigation. In line with the Integrated Marketing Communication Theory (Kitchen & Schulz, 1999), the presented analysis provides a strategic framework for developing and implementing effective and efficient negotiation strategies that are result-oriented and data-driven.

7 Implications

7.1 Managerial Implications

For managers in charge of selecting and deploying communication technologies for negotiations, the results of this study have important implications for refining performance (Diamond & Weiss, 2013; Lewicki & Litterer, 1985; Raiffa et al., 2002; Shell, 2006; Watkins, 2014). Under agreement procedures, in which negotiators try to find a solution according to their preference, a targeted use of negotiation media promises success (Adair & Loewenstein, 2013; De Dreu & Van Kleef, 2004a, 2004b; Dwyer, 2018; Salacuse, 2005). Along these lines, results imply that subjects beyond normative characteristics of tasks and technology influence negotiators' communication performance. Accordingly, negotiating managers require a comprehensive understanding of the tangible impact of media channels on a bargaining development. To grasp this channel impact on the bargaining development, it is initially essential to understand the underlying factors for channel choice, identified in this study as geographic, gender, industry and hierarchical causes.

Geographic Stimulus on Media Choice

The large international sample size provides needful insights into geographic differences, which occur during the heterogenous selection of negotiation channels. Geographically, the highest number of bargaining channels is employed in South America (3.64 mean value), and the least channels are active in North America (2.92 mean value). Typically, this could be caused by a global north-south divide (Kiely, 2016) in terms of a regulating and legislative framework towards negotiations and could also be tied to cultural aspects (Brett, 2014; Graham et al., 1994), which requires further investigation. Data reveals that North Americans and Western Europeans negotiate more frequently with classical negotiation media - email and video, whereas South Americans and Africans favorably negotiate with new negotiating channels.

South Americans, and Africans more frequently and Western Europeans less frequently select messaging compared to an equal distribution. Negotiators in South America even surpass the expectancy target for messaging in negotiations by 190%, followed by African negotiators exceeding the expectancy target for messaging in negotiations by 115%. Conclusively, using messaging for business purposes is still an ongoing legal, regulatory, and ethical debate (Mars et al., 2019) managed individually by organizations globally.

Africans use less video and email than expected in negotiations, whereas Western European negotiators use disproportionately more video calls than expected. This phenomenon could be linked to the COVID pandemic aftereffects. Surprisingly, North Americans use far less face-to-face in negotiations than expected. Under the described circumstances, managers must be cautious that local legislation, company requirements, and cultural aspects often determine the type and quantity of channels that they use.

Gender Stimulus on Media Choice

Study results demonstrate that solid gender differences occur in the choice of negotiation style when using multiple media channels. Accordingly, the negotiation style of men differs from females. Men use slightly more negotiation channels (3.2 mean value) than females (2.9 mean value) and use the phone for negotiations more often than women. Male negotiators also habitually opt for face-to-face negotiations than females. This female media withdrawal or avoidance behavior can be associated with a third category of differentiating bargaining acts. Styles of bargaining can be divided into integrative (cooperative), distributive (antagonistic), and avoidance (withdrawal) (Bednar & Curington, 1983; Harris, 1996; Putnam & Jones, 1982; Sillars, 1980). The application of these three behavioral bargaining categories varies with growing inclusiveness and competitiveness (Donohue & Roberto, 1996; Mannix et al., 1995; Sillars et al., 1982; Weingart et al., 1993). Respectively, the workplace environment needs to tolerate different negotiation styles (Moran & Ritov, 2007). This finding has implications on theory and practice as empirical data within the scope of this study proves that one-negotiation-style-fits-all is redundant.

Industry & Organizational Size Stimulus on Media Choice

For cross-industry negotiations, channel use can be typical to an industry and determined by the size of an organization. Specifically, the petrochemical and oil industries use more email in negotiations than statistically expected. This phenomenon could be linked to a convenience aspect induced by culturally cultivated email negotiations. Opposingly, less email than expected is employed by the automotive (-15%) and pharmaceutical (-23%) industries, possibly encouraged by the maturity of negotiation relationships with assorted suppliers entailing other media routines.

Empirical analysis of this research shows channel use in negotiations is shaped by the company's size. On the grounds of that, companies smaller in size quote using messaging for negotiations more often, potentially due to more liberal regulation prescribed over a company-wide negotiation conduct. The use of traditional audio and video conferencing is cited by larger companies may relate to stricter company regulations regarding the conduct of bargaining. Importantly, this study's analysis reveals that the greater the organization's size, the more advanced the effectiveness of the negotiation. Overall, this advanced effectiveness can be linked to a plethora of root causes, namely established negotiation processes, more experienced negotiators, and long-lasting supplier relationships of larger organizations.

Hierarchical Stimulus on Media Choice

Focusing on individual media use is necessary to comprehend successful communication, as a concentration on media capabilities does not yield the expected scientific results (Walther, 1992). Nearly all professional tasks include negotiations to a certain degree, which results in higher-ranking managers in organizations using significantly more bargaining channels than managers of lower ranks. Essentially, most channels (3.5 mean value) are utilized by the department head, which aligns with Thompson (2002) that leadership focuses up to 50% of its capacity on negotiating tasks. Managing directors and divisional managers use an equal amount of negotiation channels (3.2 mean value), and the least negotiation channels are deployed by team leaders (2.8 mean value), followed by managers without a team (2.9 mean value). Specifically, department heads use double the amount of audio conferencing in negotiations, as individual negotiation performance, in this case through audio-conferencing, determines professional success. Higher-ranking managers are more engaged in negotiations and, consequently, require more channels to fulfill negotiation demands. Additionally, among other reasons, department heads often conduct internal negotiations within organizations with audio as a sufficient medium to transport a message due to the familiarity with the negotiation partner(s) and organization. Even so, analysis reveals those respondents acting as division head or team lead display fewer effective negotiations than negotiators in other positions. This finding is opposed to results on the effectiveness of hierarchy negotiations in which higher management levels negotiate more effectively (Kyl-Heku & Buss, 1996; Lund et al., 2007), which in this study does not show a robustness across cultures and contexts of multi-channel negotiations.

Conditioning Negotiators to Manage Multiple Channels

Half of the negotiators in this study selected channels to stimulate the negotiation's process output. Creating a dynamic and challenging negotiation with multiple channels (Putnam & Holmer, 1992) is intentional and purposeful. This result supports the findings of Herbst et al. (2011) that negotiation research needs to drill into the depths of a successful negotiation process management. To such a degree, thoroughly examining the negotiation context is imperative for any negotiating manager to select appropriate negotiation channels. In line with Mayer & Voeth (2022), this study's results illustrate an intense desire among negotiators to manage the negotiation process actively. This negotiation management commences with a controlled planting and execution of a negotiation strategy and tactics (Voeth & Herbst, 2015) through channel selection and channel orchestration. Potentially, specific channels are selected with a documentation capacity if transparency over the exchange is an individual or organizational requirement. As negotiations can be lengthy in duration, negotiators in this survey reported using channels for intentional time management during the negotiation, in line with Voeth et al. (2020). These results underline the actual steering capacity of an insightful channel choice regarding managing time. As personal convenience is an ideal platform to manage individual negotiation performance, individual negotiator demand for convenience regarding a specific channel comfort is satisfied by a customized channel choice. However, Nadler (2004) urges cautiousness in relying on, for example, the convenience of email as a negotiation tool, as channel choice must match the situational context.

Respondents select channels for transparency on behavior to grasp and interpret the opponent's body language, nonverbal cues, reactions, facial expressions, and emotions and, with this, counteract the bargaining episodes more successfully. This empirical evidence from negotiation practice mirrors the importance of social-psychological measures for assessing negotiation (Thompson, 1990). A managerial expansion could entail coaching motivational and cognitive briefs on individual negotiation behavior. However, this finding does not support the theory of the Barrier Effect by Lewis and Fry (1977) and Carnevale et al. (1981), in which direct eye contact communicates dominance and visual contact suppresses collaboration along the bargaining process (Pruitt, 1981). These results confirm that the negotiator actively selects appropriate channels and steers the negotiation strategically in correspondence with the current behavioral negotiation setting that is in place.

Companies need to be aware that the trend is steering away from single-channel negotiations to using two to four negotiation channels, as stated by 75% of respondents with 539 counts. Negotiators favor and apply diverse media (Neslin & Shankar, 2009). On account of this, it is imperative to train the workforce to negotiate with multiple channels. Significant correlations substantiate that the older and more experienced the negotiator, the more channels are used. If negotiation experience is prevalent, then negotiators tend to use approximately one additional channel (3.1 mean value) as opposed to negotiators with no experience (2.2 mean value). This finding implies that using several negotiation channels can be trained as, based on results from data, it requires age and experience to deal with this challenge. Remarkably, older negotiators use significantly more video conferencing than the younger negotiator generation. Moreover, more experienced negotiators use the triad of video conferencing, phone calls, and email more regularly than negotiators with less experience. In line with Bandura (1997, 2001) and Volkema et al. (2013), experience increases confidence in successful target achievements. The more solid this manifestation in successful negotiations becomes, the higher is the encouragement to expand on channel negotiations.

Empirical data confirms that foresight is imperative in training managers (Nadler et al., 2003) to steer the negotiation process strategically (Adair & Brett, 2005) by selecting spot-on channels at a precise time. This result aligns with Malhotra & Bazerman (2007, p. 4) to "structure and execute negotiations strategically and systematically." Conclusively, it is recommended to expand the widely used Negotiation Management Toolbox by Voeth & Herbst (2015) with imperative channel preparatory and channel management phases for an effective and efficient steering of a multi-channel negotiation (see Figure 19).

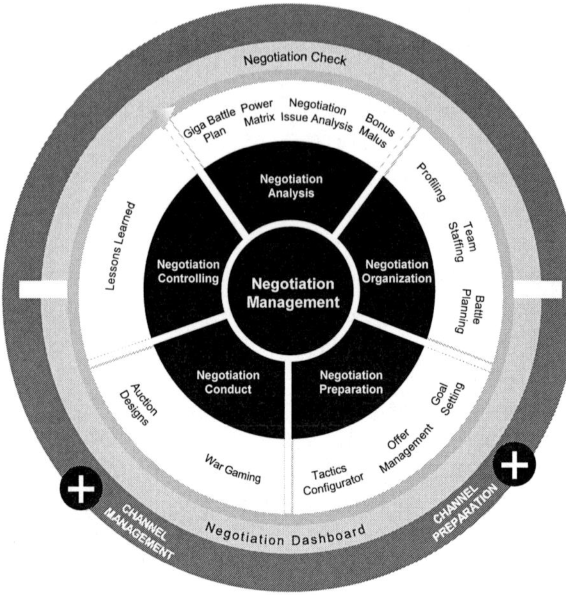

Figure 19. Channel Add-Ons for the Negotiation Management Toolbox

Performance Enhancement through Behavior in Channel Negotiations

Communication behavior can impact the success of a negotiation, which leads manager's interest in the effectiveness and efficiency of negotiation behavior (Carnevale & Pruitt, 1992; Gelfand & Dyer, 2010; Neale & Bazerman, 1992). Negotiators are often not aware of positive or negative behavioral effects (Adair et al., 2001; Neale & Bazerman, 1991; Pruitt & Carnevale, 1993; Thompson & Hastie, 1990; Tinsley & Brett, 2001) during a multi-channel negotiation process and outcome. In this context, behavioral dimensions are a lever that directly impact up to 30% of negotiation effectiveness. The selected behavioral dimensions even influence negotiation efficiency by 24%. These numerical values derived from statistical analysis reveal a massive bargaining growth potential as an increase of negotiation performance has a considerable impact on profit (Voeth & Herbst, 2015). Specifically, self-efficacy, agility, and behavioral control improve negotiation effectiveness and efficiency considerably, leaving ample opportunities for managers to dive into manifold coaching measures (Bazerman et al., 1985; ElShenawy, 2010; Herbst et al., 2019; Voeth et al., 2015) on these specific negotiation behaviors. Unexpectedly, within the context of behavioral dimensions, complexity directly impacts effectiveness and efficiency. It also acts as an indirect strengthening amplifier on behavioral control and effectiveness and efficiency. These findings support the review of Koch & Shepperd (2004) that self-complexity corresponds to effective coping with adverse events (Dixon & Baumeister, 1991; Linville, 1985, 1987). However, complexity does weaken the weight of self-efficacy on efficiency. Thus, self-efficacy only transforms into negotiation leadership in a reduced, simplistic environment. Interestingly, performance risk buffers the effect of synchronicity on efficiency, and cognitive resource demands indirectly reduce the effect of synchronicity on effectiveness - confirming two of the originally formulated hypotheses. This conclusion supports the assumption that synchronicity needs numerous sub-processes and time to evolve (Geiger, 2020) as a prevailing risk of performance and engaging high cognitive capacities in a multi-channel setting blocks its emergence.

Channel Quantity Impacts Negotiation Behavior & Efficiency

The usage of channels impacts and creates a change in negotiation behavior. With increased channels, the importance of behavior in steering a multi-channel negotiation increases. In this section, behavioral access on efficiency is first discussed, followed by details on drivers for effectiveness. The strongest direct drivers on negotiation efficiency are self-efficacy and behavioral control. As Judge et al. (2002) elaborate, self-efficacy is a powerful manifestation (Vroom, 1964), which materializes into success (Volkema et al., 2013) and explains the most variance mentioned by Kirk & Brown (2003) in negotiation efficiency. Throughout channel usage, self-efficacy continues to provide a stable and positive impact on effectiveness regardless of whether negotiators deal with a single or even four or more channels. The manifestations of self-efficacy paired with the negotiator's control over objects and processes is incremental to achievement. Results confirm that behavioral control regulates performance (Bandura, 1977; Lim & Weissmann, 2021; Terry & O'Leary, 1995) efficiently.

Moreover, enhanced agility is a driver for efficient conduct in negotiations. Along these lines, negotiator agility is a flexible and adaptive thinking and behavior, especially in response to changing circumstances. It is the capacity to absorb novel abilities quickly, adapt to new situations, and respond to challenges with resilience. Specifically, an agile negotiation occurs through in-time access to relevant information via multiple negotiation channels, which is interpreted and applied. As Loewenstein & Thompson (2006, p.77) state, “everyone negotiates, and everyone could ... learn to negotiate better.” The study is relevant to negotiation training to condition a self-effective, agile, and controlled negotiation behavior in multi-channel negotiations.

Complexity is a stable direct driver for efficiency. As data records of this study confirm, it corresponds to effective coping with adverse events (Dixon & Baumeister, 1991; Koch & Shepperd, 2004; Linville, 1985, 1987). It acts favorably on behavioral control and efficiency yet minimizes agility and efficiency when using double the number of channels. Ultimately, the more the contextual complexity, the stronger the negotiator's control over objects and processes. Conclusively, negotiation sciences necessitate to further explore a cut-off point between favorable vs. adverse complexity effects. In summary, complexity is not uniquely harmful in impact in a multi-channel environment.

Volatile results of an anticipated performance risk (Mitchell & Greatorex, 1993; Stone & Gronhaug, 1993) on negotiation efficiency suggest the subjective inability of negotiators to perform in an environment of uncertainty (Ahuja & Carley, 1998; Daft & Macintosh, 1981; Edström & Galbraith, 1977; Galbraith, 1973; Tushman & Nadler, 1978). Thus, performance risk is a direct buffer to efficiency when applying five or more channels. Lutz & Reilly (1974) explain how a process performs differently than expected, making the inability to perform in this study closely linked to five and more channels. Throughout channel usage, performance risk diminishes synchronicity effects on efficiency. Unpredictably, with three channels, performance risk acts favorably on self-efficacy and efficiency. This result could be linked to the well-trusted stability created by three channels, as confirmed by multiple study participants. Consequently, more than three channels lead to an unfavorable impact of performance risk on the negotiation outcome.

Centered around work of Wilcox et al. (2011), cognitive resource demands - in the context of this study - is the level of thinking and attention a negotiator devotes to a multi-channel negotiation. While on the one hand, cognitive resource demands reduce synchronicity and self-efficacy effects on efficiency in high-channel negotiations. On the other hand, cognitive resource demands, at the same time, also improve agile effects on efficiency using four channels. As depicted by Herbst et al. (2017), the negotiator's cognition processes a variety of data on opponents and objects, resulting in overloaded cognitive processing abilities (Geiger & Hüffmeier, 2020), reducing synchronicity and self-efficacy effects on efficiency in high channel negotiations. In high-channel negotiations, unlocking content cognitively to strategically plan a counterstrike (Lee, 1994; Petty & Cacioppo, 1986) is impossible. Thus, creating synchronicity in encoding negotiation messages through manifold channels is a natural impasse. Based on research findings of Miranda & Saunders (2003) and Sitkin et al. (1992), the positive effect of cognitive resource demands can be a cause of specific media in this high-channel environment, which enables the transfer and processing of information to create meaning. Naturally, this phenomenon of the interrelation of cognitive resource demands and agility in a favorable media landscape requires further investigation.

Reaching the Performance Optimum in Channel Negotiations

In a multi-channel bargaining environment, the optimum behavioral steering capacity is reached at three channels, which confirms that multiple communication media transmit a message more effectively and efficiently to a target group (Kunz & Kunz, 2014a, 2014b). Negotiators "must effectively manage multiple channels to increase cross-channel synergy and reduce channel conflicts" (Wang et al., 2015, p. 390). Herein, especially a self-effective behavior, with specifically an overall most potent behavioral effect, shapes effective and efficient negotiations decisively towards a positive outcome. In this sense, for behavior modification, self-efficacy is required as it reflects people's confidence and motivation to execute performance with perseverance (Bandura, 1977, 1982; Kirk & Brown, 2003; Lim & Weissmann, 2021). Consequently, the weight of self-efficacy targeting success is a prime area for further scientific and practical explorations. Conversely, complexity proves surprisingly positive in its impact on the negotiation outcome with "the implication that we need to find ways to embrace rather than reject complexity" (Weisberg et al., 2013, p. 37). It acts as a strong driver for better negotiation effectiveness and efficiency. Furthermore, throughout the use of three to five and more channels, agility yields favorable effects on efficiency. This fused flexibility structure thrives under heterogeneous and ambiguous settings (Charbonnier-Voirin, 2011). To build agility, individuals need sufficient information on unexpected conditions and capabilities for processing (Alavi et al., 2014; Cai et al., 2018). High channel quantities trigger flexible and adaptive thinking and behavior, especially in response to changing circumstances (De Meuse et al., 2010).

As the optimum of three channels is exceeded, effective steering of the multichannel negotiation with the selected behavioral dimensions decreases when using more than four negotiation channels. In line with Wang et al. (2015, p. 390), in this high-channel environment, negotiators within this study's scope cannot "reduce channel conflicts." Nevertheless, only the positive effect of behavioral control increases with a growing number of channels. While using four channels, self-efficacy and behavioral control are substantial pillars in maneuvering the negotiation into success with high efficiency and effectiveness values.

Behavioral control displays a stable effect on four or more channels and positively augments effectiveness. It is prevalent if the behavior is under the negotiator's control (Lim & Weissmann, 2021; Terry & O'Leary, 1995). In line with theory on observed behavioral control (Ajzen, 1991, 2012; Fishbein & Ajzen, 2010; Trenz, 2015) and the Planned Behavior Theory (Lim & Weissmann, 2021), negotiators intentionally expand control over objects and processes in advanced channel negotiations. More control is demanded of negotiators when using five or more negotiation channels, which positively impacts the overall efficiency and effectiveness of the negotiation. The behavioral control mechanism depends on information, experiences, and anticipated assistance and impediments (Lim, 2003). It increases with perceived greater resources and opportunities and when negotiators anticipate fewer obstacles and impediments. Elements of behavioral control can be intrinsic in the form of abilities, capacities, perseverance, and obligations or induced through external factors such as duration, prospects, and reliance (Ajzen, 1985, 1988; Sparks et al., 2006). Additionally, self-effective behavior proves vital for success in effectiveness within this multifold channel landscape. Nevertheless, with increased cognitive demands in five or more channel negotiations, the self-effective behavior lessens, which leads to an overall loss in efficiency. The negotiator fails to unlock cognitive content, which as a result, inhibits self-efficacy from blossoming. As timely cognitive processing hits its limit in manifold channels, this negatively impacts confidence in self-effective and efficient performance. Herbst et al. (2017) define this phenomenon in negotiations as overloaded cognitive processing abilities (Geiger & Hüffmeier, 2020).

Lastly, Media Synchronicity Theory matches the communicator's synchronicity with media synchronicity through "conveyance and convergence processes" (Dennis et al., 2008; Parlamis & Geiger, 2014, p. 362). Communication, especially negotiation, consists of subprocesses in which the amount of each conveyance and convergence differs (Geiger & Laubert, 2018). Derived from Arthi (2009) and Dennis et al. (2008, 2009), synchronicity in multi-channel negotiations is defined as a coordinated, synchronous behavior of the negotiation parties - caused by using diverse negotiation channels - with a shared goal to create a win-win situation, that can lead to an integrative negotiation. Conspicuously, there is no positive impact of synchronicity, nor an indirect enhancement of synchronicity on the outcome variables, making it irrelevant in impact. It could be that synchronicity is a rather abstract concept making it difficult for study participants to comprehend. This theoretical concept might be comprehensible on paper. However, until to date it has been less empirically recorded in negotiation practice. Due to its release date, the concept of Media Synchronicity Theory is less tested by the research community than Media Richness Theory or Grounding Theory. Furthermore, the prime intention of negotiators is only sometimes integrative, as opposed to the ideals of synchronicity theory. Doubts emerge if a coordinated and synchronous behavior (Dennis et al., 2008) of negotiating opponents is even realistic and upholds the nature of diverging interests and competitive actions in negotiations. Realistically, an efficacious matching of negotiation channels with the negotiation content to establish synchronicity does not take place to impact negotiation effectiveness and efficiency positively.

Insights into the impact of behavioral dimensions in multi-channel negotiations serve as an essential orientation. So far, no comparable international study covering 733 international negotiation experts exists. The findings of this quantitative study specify a foundation for managerial strategies and the corresponding required profound judgment regarding the challenges in global negotiations. Conclusively, this quantitative study reveals a direction of strategic behavior in multi-channel negotiations. The implications feature potential strategic approaches to administer multi-channel negotiations effectively and efficiently.

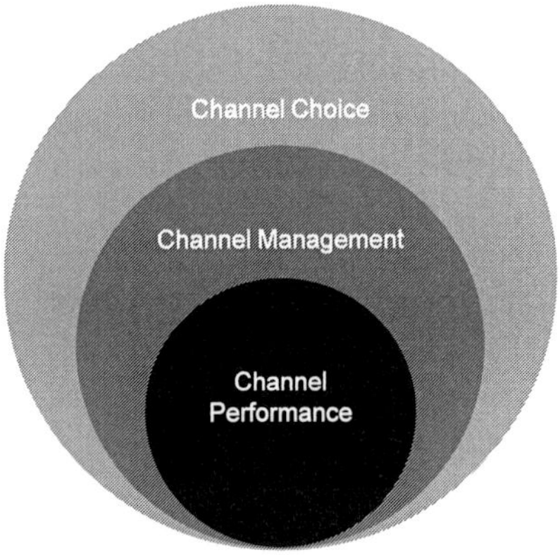

Figure 20. The Success Triade of Channel Management

7.2 Theoretical Implications

It is necessary to review communication theories of a strategic nature that center around media and assignment' features and advocate a solitary channel or channel arrangement for communicative interactions (Geiger, 2020) to comprehend the concept of multi-channel negotiations. Strategic communication theories analyze media characteristics and tasks, whereas social-psychological perspectives focus on the communicator–medium interaction (Geiger, 2020). A communication medium can aid or distract social processes essential to communicators. In due course, social-psychological perspectives will complement the theoretical grounding work for multi-channel negotiations. Most of the relevant communication theory selected for this scientific work concerns communication in the broad sense and does not focus on the context of negotiations. Conversely, the outcomes of this analysis are coherent with the Integrated Marketing Communication Theory (Kitchen & Schultz, 1999). In line with this theory, the findings of this study demonstrate that channels work synergistically throughout different negotiation stages and that targeted use of negotiation media promises success (Adair & Loewenstein, 2013; De Dreu & Van Kleef, 2004a, 2004b; Dwyer, 2018; Salacuse, 2005).

Differentiation in Conveyance and Convergence is Substantial

Media Synchronicity Theory matches the communicator's synchronicity with media synchronicity through "conveyance and convergence processes" (Dennis et al., 2008; Parlamis & Geiger, 2014, p. 362). Following this concept, synchronicity - caused by using diverse negotiation channels - is a coordinated, synchronous behavior of negotiation parties with a shared goal to create a win-win situation, which can lead to an integrative negotiation (Arthi, 2009; Dennis et al., 2008, 2009). Communication, especially negotiation, consists of subprocesses in which the amount of each conveyance and convergence differs (Geiger & Laubert, 2018). Each subprocess in a negotiation requires a different channel for successful conveyance and convergence. The dominant claim among global negotiators is that the heart of the negotiation, which is the issue discussion, is still very traditionally conducted synchronously via face-to-face or video conferencing. This communication conduct aligns with Synchronicity Theory's concept that information is discussed and interpreted in a face-to-face or video conferencing convergence process. This recorded negotiation behavior is an attempt to efficiently match negotiation channels with the negotiation content to impact negotiation effectiveness and efficiency positively. With an unanimous resolution, negotiators manage processes surrounding the central part of the negotiation with asynchronous email. Thus, information is transported to the receiving end through an electronic conveyance process to construct a cognitive understanding of the content. Danaher & Rossiter (2011) urge email is easier to reject, generally rated more negatively and only sometimes translates into more effective negotiation outcomes (Geiger, 2020). On that account, negotiators could tend to be overly optimistic concerning the overall success rate of email.

However, scientific findings underline that email is more trustworthy and "convenient, acceptable, and appropriate" (Danaher & Rossiter, 2011, p. 18). Thereupon, media channels like chat, messaging, and audio conferencing play a minor role in negotiation stages, and negotiators still prefer to rely on classical negotiation media. Opposed to Mennecke (2000), single communication technologies prove adequate for phases of a single task.

In line with Synchronicity Theory, the differentiation in conveyance and convergence processes in a negotiation is substantial. Along these lines, the appeal is not to use identical channels for information sharing and consensus building. The most critical negotiation channel for sharing information is face-to-face. Video and audio conferencing, email, and phone calls are equally well perceived for information sharing. Essential consensus-building channels are face-to-face, video conferencing, and phone calls. The main difference in information sharing is that respondents select a phone call instead of email to arrive at a mutual decision. Subsequently, it can be confirmed that all three channels display a visual or audio function essential for decision-making. In line with information sharing, chat, and messaging channels are the least useful for consensus building.

Absence of Synchronicity Impacts on Effectiveness and Efficiency

Despite the above findings, outcomes of this analysis do not evidence a positive weight of synchronicity on negotiation effectiveness and efficiency. For this reason, the efficacious matching of negotiation channels with content does not automatically establish synchronicity. Nevertheless, most respondents (80%) agree that using multiple channels effectively enhances the negotiation relationship to gain more information at the appropriate time. To advance this perspective, nearly 80% of negotiators agree or strongly agree that multi-channel negotiations are the appropriate media mix for creating a shared understanding. Lastly, over 80% of negotiators find that multi-channels fit well with the demand to deliver information. According to the respondents of this study, multi-channel negotiations fulfill the need for information sharing and creating a shared understanding. Notwithstanding, synchronicity's sought-after positive statistical effects do not materialize in negotiations within this study's scope. Thus, this concept remains part of negotiation theory.

Desired Co-Construction of a Shared Understanding Does Not Materialize

The needful conveyance as part of Synchronicity Theory – an information exchange - is a solid negotiation component. Often strategy and tactics determine the amount and timing of information shared during the bargaining process. In line with Voeth et al. (2020), information processing occurs under time restrictions. Multi-channel bargaining is leveraged to deliver information in the best possible way. Henceforth, information sharing in negotiations is a delicate topic that needs differentiated approaches and analysis. Furthermore, synchronicity is about a shared understanding. Ultimately, a negotiation will necessitate conveyance and convergence in different proportions and durations depending on the type of negotiation or individual. When negotiators differ in perspective, more intellectual capacities are needed than for information sharing. Sitkin et al. (1992) rightly state that negotiation parties only sometimes co-construct a shared understanding. Early negotiation science differentiates between integrative and distributive bargaining behavior (Bednar & Curington, 1983). Consequentially, it is acceptable for negotiation parties to maintain their individual vs. mutual understanding and a corresponding goal in a time-limited (Voeth et al., 2020) and foremost competitive exchange process. Negotiation research, in fact, strongly supports that bargaining, in general, is managed with a competitive attitude (Neale & Bazerman, 1992; Pruitt & Carnevale, 1993). Negotiators can either agree on the objects, process, or outcome or settle on a non-conformist ground (Lind & Zmud, 1991). In this manner, negotiation parties follow their goal orientation unless integrative negotiations are the target. Only through integrative negotiations does this mindset transform into a mutual understanding or a win-win negotiation. Miranda & Saunders (2003), Rogers (1986), and Te'eni (2001) pledge that the communicative intention is to settle on common ground. This research strongly suggests that a mutual understanding in negotiations is a benefit. However, in the context of competition, it is not always desired or mandatory. Consequently, the interaction might align the negotiator's comprehension, but this alignment is frequently absent. In this way, a multi-channel negotiation is a format to create a shared understanding among negotiators. Developing a shared meaning is a rather abstract concept when it comes to buyer-supplier relationships which are unemotionally number-oriented. The question is whether, in this relationship, the development of meaning is compulsory. One approach is that this depends on the buyer-supplier relationship and how sustainable it is. Buyers often share number-oriented information and have minor concerns about how others may interpret this information for the transaction. Thus, attaining synchronicity often depends on the context of the negotiation. Opposed to the assumption of Lind & Zmund (1991), the purpose of negotiators is rarely to settle on common grounds. Instead, somewhat negotiators agree or partially agree on objectives of the negotiation. The nature of most negotiations consists of diverging understandings, interests, and opposing targets. It follows that the ideal of an individually shaped and socially developed phenomenon of synchronicity does not unfailingly emerge.

Due to its release date, the concept of Media Synchronicity Theory is less tested by the research community than Media Richness Theory or Grounding Theory. To further challenge and validate Synchronicity Theory, as well as the findings of this study on the behavioral dimension of synchronicity in negotiations, more empirical studies on the impact of synchronicity are of utmost importance. The following chapter highlights further research requirements on the topic of multi-channel negotiations.

8 Limitations & Prospective Research Opportunities

8.1 Limitations

Limitations of this study apply in terms of its comprehensive expansion on contextual factors, the processual aspect of the duration of the negotiation, and inclusive analysis of behavioral traits. The study is limited in its geographic reach. It presents a snapshot in time of an evolving media landscape, it delivers an isolated view of a single negotiation entity instead of negotiation teams or even the interplay of dyadic parties, and it does not consider situational aspects influencing the type of negotiation.

Contextual Limitations

Numerous international studies on negotiation practice (Adair & Brett, 2005; Liu et al., 2010; Lügger et al., 2015) emphasize the importance of not only relying on intracultural negotiation perspectives, which are negotiations between national counterparts (Adler et al., 1987; Metcalf et al., 2006). For this reason, 733 international negotiation managers evaluate the effectiveness and efficiency of negotiation behavior in multi-channel negotiations. Broadening the empirical research to an international setting (Liu et al., 2012; Salmon et al., 2013) follows the assumption that intracultural negotiation cannot be identically transferred to intercultural negotiations (Adair & Graham, 1989). A sizeable international sample provides necessary insights into heterogeneous cultures and corresponding diverse negotiation behavior (Brett, 2014; Graham et al., 1994). However, over 60% of the survey participants (394) originate from Europe, divided into 37 Eastern Europeans, 40 Northern Europeans, 17 Southern Europeans, and 292 Western Europeans. The group of Western Europeans consists of predominantly German nationals (257 counts). In conclusion, half of the study results are opinions of Europeans, and of these even a third are the beliefs and judgments of German nationals or negotiators residing in Germany. This finding implies the need to consider this solid regional, national, and cultural representation in the sample and avoid a generic transfer of results to any global context.

Industrial negotiations are often conducted in teams and are imperative in influencing a company's success. However, team negotiations lack attention in general sciences - in general, business administration, industrial goods marketing research, group or team research, or in negotiation sciences (Barisch, 2011). Media theories must carefully consider the entire group process, not simply task execution activities. The study of Voeth et al. (2022) reveals that team negotiations gain importance with the growing sizes of negotiation teams. An essential limitation of the empirical findings processed in this work is that empirical evidence from this questionnaire derives from an individual negotiator and not a team of negotiators. In the scope of this study, it is unclear if the individual respondents might have been involved in a group negotiation or a dyad in their last significant negotiation, as this question was not part of the survey. In further studies, this background information is essential for integration in a survey catalogue as it has implications on the subsequent data analysis.

External factors and the level of digitization in business will resume transforming negotiations, and with this, new forms and channels of negotiation will continue to emerge. Thereupon, the choice of negotiation media tomorrow will continue to be vast compared to decades ago (Agndal et al., 2017). Subsequently, this study only exposes a snapshot of media usage in negotiations. To such a degree, the media landscape for negotiations is dynamic and evolving.

As bargaining entails a situational context of competition and cooperation (Adam & Brett, 2015; Galinsky & Schweitzer, 2015; Halevy & Phillips, 2015; Pruitt & Carnevale, 1993; Yip & Schweinsberg, 2017), using multiple channels depends on the negotiation context. McGinn and Croson (2004) urge caution in seeing only the positive aspects of multiple negotiation channels. For example, on the one hand, a shared understanding regarding the negotiation objects is, in some instances, favorable. On the other hand, minimized social awareness through fewer channels in emotionally charged negotiations (e.g., war) is of benefit, according to Poole & DeSanctis (1992). As such, in the case of maximum profit as an individual or collective negotiation goal, low social awareness through fewer channels is advantageous to the negotiation. Thus, cautious judgment is advised in channel type and quantity selection, depending on the context and topic of the negotiation.

Furthermore, information on negotiation objects is another relevant factor requiring consideration. A negotiation object can contain single or multiple negotiation issues, such as price, quality, and timing – and with this determines a minimal or maximum scope of a negotiation (Jennings et al., 1998). This setting can entail high complexity, with negotiators “reaching multiple agreements over multiple objects” with multiple parties (Mansour & Kowalczyk, 2014, p. 46). A measurement of the success of interconnected agreements about objects, channels, and behavior is of interest. This study did not specifically collect information on negotiation objects and corresponding intrinsic issues. Relating the negotiation objects to the channel choice in reaching an effective or efficient outcome is another essential aspect to be considered for further studies. Particularly, changing the structure of negotiation objects by adding content concerning issues (Jennings et al., 1998) could yield potential insights into strategic behavioral-channel interceptions. This change implies evolving negotiation strategies for the negotiation setting of compounded objects (Mansour & Kowalczyk, 2014).

Processual Limitations

The duration of a negotiation is incremental for its success as prolonging it results in costs and uncertainty (Baarslag et al., 2015). Galin et al. (2007) compare the time of e-negotiations versus face-to-face negotiations and record no significant discrepancies. Analyzing negotiation time in relation to negotiation media might indicate paths to shorten lengthy trade negotiations (Lechner & Wüthrich, 2018; Simonelli, 2011). In contrast, this survey does not explore a multi-channel negotiation's duration in hours, days, weeks, or even months. The measured duration of a multi-channel negotiation might reveal insights into channel use and behavioral efficiency.

This study assesses the impact of four selected behavioral dimensions relevant to multi-channel negotiations. The findings are established from the selected behavioral dimension and do not universally apply to negotiation behavior. A differentiated approach to diverse forms of negotiation behavior is essential and implies the necessity for further research. Despite these limitations, this scientific research produces a solid foundation for future academic studies on effective and efficient behavior in multi-channel negotiations, which is a critical issue for negotiation research.

8.2 Prospective Research Opportunities

Transparency on Contextual Implications

In a transparent contextual framework, the effects of media combinations on bargaining behavior, the negotiation process, and the outcome need to be further explored holistically. On that account, in the negotiation context, a negotiation's nature and setting determine the negotiation channel's selection. The deductions of this study are grounded in contemporary media. Conversely, with advancements in technology, it is essential to persistently investigate the dynamic channel landscape to validate research and further enhance scientific findings on the research topic of multi-channel negotiations. Empirical evidence exists on correlations between the negotiation channels in multi-channel negotiations; nonetheless, these results altercate with the modification of the negotiation context. For this, it is vital to consider the negotiation context in any future research.

The Social Information Processing Theory (Walther, 1992, 1994) and the concept of communication orientation (Swaab et al., 2012) partially unpuzzle the interrelation of context, negotiator, and opponent in multi-channel negotiations. Accordingly, it is imperative to include contextual factors and an in-depth focus on the negotiator and the opponent in the analysis. This interrelation of context and negotiation behavior between parties in multi-channel negotiations requires comprehensive studies. As a proposal, this study can be conducted in pairs with the buyer and corresponding seller to see how this influences the choice of negotiation channel, negotiation behavior, and outcome. Swaab et al. (2012) suggest that multi-channel negotiation – in the form of audio, visual, and synchronicity – reinforces the positive negotiation effect for neutral negotiators, mitigates non-cooperative negotiation behavior, and does not interfere with cooperative negotiators achieving joint profit. In accordance, further exploration requires a focus on the negotiator's attitude – personal behavior or character traits – and the choice of negotiation channel as well as the resulting outcome of the negotiation. Potential investigations should test the impact of different channels on task participation and the social psychological process surrounding it more closely.

Transparency on Processual Implications

Difficulties emerge in defining a negotiation process, observing, analyzing, and evaluating it (Zartman, 1988). "Understanding what constitutes the essence and the reality of the negotiation process has been a major concern of theoreticians" (Dupont & Faure, 1991, p. 67) and, against the backdrop of multi-channel negotiations, implies endless opportunities for further exploration in this field. Accordingly, practitioners within this study specifically requested to know a multi-channel negotiation's accuracy and distortion effect along the negotiation process. For this, comparisons of different forms of multi-channel negotiations can take place, for instance, in an experiment to comprehend the influence on options for an agreement. Furthermore, for deep analysis, each negotiation is divided into subprocesses with corresponding tasks and matching channels. In that event, how each negotiation task with respect to its objects, target achievement, and uncertainty influences the interplay of multi-channel negotiations needs to be explored. Furthermore, the tension of a preferred channel and its employment in what stage of the negotiation needs consideration. Ultimately, the negotiation process is of great relevance to the resulting negotiation behavior.

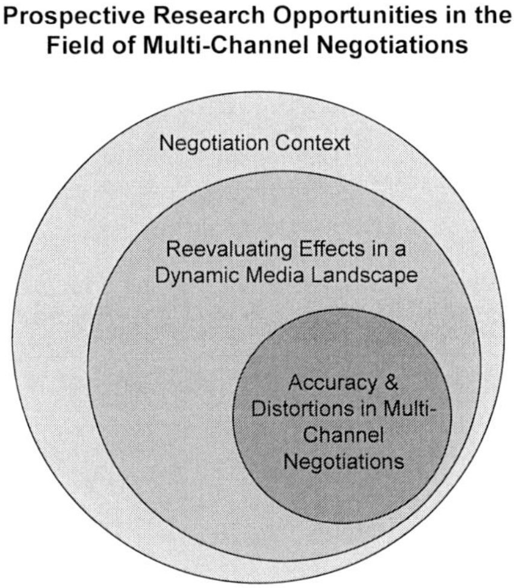

Figure 21. Prospective Multi-Channel Research Opportunities

References

Adair, J. E. (2005). How to grow leaders: The seven key principles of effective leadership development. Kogan Page Publishers.

Adair, W., Brett, J. (2005). The Negotiation Dance: Time, Culture, and Behavioral Sequences in Negotiation. Organization Science, vol. 16(1), 33-51. https://doi.org/10.1287/orsc.1040.0102

Adair, W. L., Okumura, T., & Brett, J. M. (2001). Negotiation behavior when cultures collide: The United States and Japan. Journal of Applied Psychology, 86(3), 371. https://doi.org/10.1037/0021-9010.86.3.371

Adair, W. L., & Loewenstein, J. (2013). Talking it through: Communication sequences in negotiation. Handbook of research on negotiation, 311-331. Edward Elgar Publishing. https://doi.org/10.4337/9781781005903.00022

Adam, H., & Brett, J. M. (2015). Context matters: The social effects of anger in cooperative, balanced, and competitive negotiation situations. Journal of Experimental Social Psychology, 61, 44-58.

Adler, N. J., Graham, J. L., & Gehrke, T. S. (1987). Business negotiations in Canada, Mexico, and the United States. Journal of Business Research, 15(5), 411-429. https://doi.org/10.1016/0148-2963(87)90010-5

Afflerbach, T., (2020). Theory: Challenges for Cooperation. Hybrid Virtual Teams in Shared Services Organizations: Practices to Overcome the Cooperation Problem, 9-50.
https://doi.org/10.1007/978-3-030-34300-2_2

Agndal, H. (2007). Current trends in business negotiation research. Stockholm School of Economics Research Paper, 3, 1-55.
https://swoba.hhs.se/hastba/papers/hastba2007_003.pdf

Agndal, H., Åge, L.-J., & Eklinder-Frick, J. (2017). Two decades of business negotiation research: An overview and suggestions for future studies. *Journal of Business & Industrial Marketing*, 32(4), 487–504.
https://doi.org/10.1108/JBIM-11-2015-0233

Aguinis, H., Beaty, J. C., Boik, R. J., & Pierce, C. A. (2005). Effect size and power in assessing moderating effects of categorical variables using multiple regression: a 30-year review. *Journal of applied psychology*, *90*(1), 94.
https://doi.org/10.1037/0021-9010.90.1.94

Ahuja, M. K., & Carley, K. M. (1998). Network structure in virtual organizations. *Journal of computer-mediated communication*, 3(4), JCMC343. https://doi.org/10.1111/j.1083-6101.1998.tb00079.x

Aiken, L. S., West, S. G., & Reno, R. R. (1991). *Multiple regression: Testing and interpreting interactions*. Sage.

Ajzen, I. (1985). From intentions to actions: A theory of planned behavior. In *Action control: From cognition to behavior* (pp. 11-39). Berlin, Heidelberg: Springer Berlin Heidelberg. https://doi.org/10.1007/978-3-642-69746-3_2

Ajzen, I. (1991). The theory of planned behavior. *Organizational behavior and human decision processes, 50*(2), 179-211. https://doi.org/10.1016/0749-5978(91)90020-T

Ajzen, I. (2012). Martin Fishbein's legacy: The reasoned action approach. *The Annals of the American Academy of Political and Social Science, 640*(1), 11-27. https://doi.org/10.1177/0002716211423363

Al Majzoub, K. (2023). *Modeling the factors determining a virtual team's decision making* (Doctoral Dissertation, Vilniaus Gedimino technikos universitetas).

Alavi, M., Wheeler, B. C., & Valacich, J. S. (1995). Using IT to reengineer business education: An exploratory investigation of collaborative telelearning. *MIS quarterly*, 293-312. https://doi.org/10.2307/249597

Alavi, S., Abd. Wahab, D., Muhamad, N., & Arbab Shirani, B. (2014). Organic structure and organisational learning as the main antecedents of workforce agility. *International Journal of Production Research*, *52*(21), 6273-6295. https://doi.org/10.1080/00207543.2014.919420

Albers, S., & Eggert, K. (1988). Kundennähe: Strategie oder Schlagwort? *Marketing: Zeitschrift für Forschung und Praxis*, 5-16.

Albers, S., & Hildebrandt, L. (2006). Methodische Probleme bei der Erfolgsfaktorenforschung – Messfehler, formative versus reflektive Indikatoren und die Wahl des Strukturgleichungs-Modells. *Schmalenbachs Zeitschrift für betriebswirtschaftliche Forschung*, 58, 2-33. https://doi.org/10.1007/BF03371642

Alfredson, T., & Cungu, A. (2008). *Negotiation theory and practice: A review of the literature.* Rome, Italy. Food and Agriculture Organization of the United Nations. http://www. fao. org/docs/up/easypol/555/4-5_negotiation_background_ paper_179en. pdf

Ambrose, E., Marshall, D., Fynes, B., & Lynch, D. (2008). Communication media selection in buyer-supplier relationships. *International Journal of Operations & Production Management*, 28(4), 360–379. https://doi.org/10.1108/01443570810861561

Amos, J. (1998). Transformation to agility: manufacturing in the marketplace of unanticipated change. Garland. *New York*.

Andaleeb, S. S. (1995). Dependence relations and the moderating role of trust: implications for behavioral intentions in marketing channels. *International Journal of Research in Marketing*, 12(2), 157-172. https://doi.org/10.1016/0167-8116(94)00020-O

Anderson, R. E., & Huang, W. Y. (2006). Empowering salespeople: Personal, managerial, and organizational perspectives. *Psychology & Marketing*, 23(2), 139-159. https://doi.org/10.1002/mar.20104

Armitage, C. J., & Conner, M. (2001). Efficacy of the theory of planned behaviour: A meta-analytic review. *British journal of social psychology*, *40*(4), 471-499. https://doi.org/10.1348/014466601164939

Arteta, B. M., & Giachetti, R. E. (2004). A measure of agility as the complexity of the enterprise system. *Robotics and computer-integrated manufacturing*, *20*(6), 495-503. https://doi.org/10.1016/j.rcim.2004.05.008

Arthi, B. (2009). Distributed requirements negotiations using mixed media. *International Journal of Engineering and Technology*, 1(5), 474–480. https://doi.org/10.7763/IJET.2009.V1.87

Ashbrook, C. C., & Zalba, A. R. (2021). Social media influence on diplomatic negotiation: Shifting the shape of the table. *Negotiation journal*, 37(1), 83-96. https://doi.org/10.1111/nejo.12353

Avritzer, A., Beecham, S., Britto, R., Kroll, J., Menasche, D. S., Noll, J., & Paasivaara, M. (2015, July). *Extending survivability models for global software development with media synchronicity theory* [Paper presentation]. IEEE 10th International Conference on Global Software Engineering, 23-32. IEEE, Ciudad Real, Spain. https://doi.10.1109/ICGSE.2015.29

Baarslag, T., Gerding, E. H., Aydogan, R., Schraefel, M. C. (2015). *Optimal Negotiation Decision Functions in Time-Sensitive Domains* [Paper presentation]. IEEE/WIC/ACM International Conference on Web Intelligence and Intelligent Agent Technology (WI-IAT), Singapore, 2015, 190-197. https://doi.10.1109/WI-IAT.2015.161.

Backhaus, K., & Voeth, M. (2010). *Internationales Marketing*. Stuttgart: Schäffer-Poeschel.

Backhaus, K., Erichson, B., & Weiber, R. (2015). *Fortgeschrittene multivariate Analysemethoden: eine anwendungsorientierte Einführung*. Springer-Verlag.

Backhaus, K., Erichson, B., Plinke, W., & Weiber, R. (2016). *Multivariate Analysemethoden*. Springer Berlin Heidelberg. https://doi.org/10.1007/978-3-662-46076-4

Bagozzi, R. P., & Edwards, J. R. (1998). A general approach for representing constructs in organizational research. *Organizational research methods*, *1*(1), 45-87. https://doi.org/10.1177/109442819800100104

Bagozzi, R. P., & Yi, Y. (2012). Specification, evaluation, and interpretation of structural equation models. *Journal of the academy of marketing science*, *40*, 8-34. https://doi.org/10.1007/s11747-011-0278-x

Baker, M., Hansen, T., Joiner, R., & Traum, D. (1999). The role of grounding in collaborative learning tasks. *Collaborative learning: Cognitive and computational approaches*, 31, 63.

Baltes, B. B., Dickson, M. W., Sherman, M. P., Bauer, C. C., & LaGanke, J. S. (2002). Computer-Mediated Communication and Group Decision Making: A Meta-Analysis. *Organizational Behavior and Human Decision Processes*, 87(1), 156–179. https://doi.org/10.1006/obhd.2001.2961

Bandalos, D. L., & Boehm-Kaufman, M. R. (2010). Four common misconceptions in exploratory factor analysis. *Statistical and methodological myths and urban legends,* 81-108. Routledge.

Bandura, A. (1977). Self-efficacy: toward a unifying theory of behavioral change. *Psychological review*, *84*(2), 191. https://doi.org/10.1037/0033-295X.84.2.191

Bandura, A. (1982). Self-efficacy mechanism in human agency. *American psychologist*, *37*(2), 122. https://doi.org/10.1037/0003-066X.37.2.122

Bandura, A. (1997). Behavior theory and the models of man (1974). In *Meeting of the American Psychological Association, Aug, 1974, New Orleans, LA, US; Presidential Address presented at the aforementioned meeting.* American Psychological Association.

Bandura, A. (2001). Social cognitive theory: An agentic perspective. *Annual review of psychology*, *52*(1), 1-26. https://doi.org/10.1146/annurev.psych.52.1.1

Baron, R. M., & Kenny, D. A. (1986). The moderator–mediator variable distinction in social psychological research: Conceptual, strategic, and statistical considerations. *Journal of Personality and Social Psychology*, 51(6), 1173-82.

Baron, N. S. (2013). Instant messaging. *Pragmatics of computer-mediated communication*, 9. https://doi.org/10.1515/9783110214468

Barisch, S. (2011). *Optimierung von Verhandlungsteams: Der Einflussfaktor Hierarchie (Doctoral Dissertation)*. Cumulative thesis in Negotiation Management. Publisher Springer-Verlag.

Barkhi, R., Jacob, V. S. F., & Pirkul, H. (1999). An experimental analysis of face to face versus computer mediated communication channels. *Group decision and negotiation*, *8*, 325-347. https://doi.org/10.1023/A:1008621423120

Barkhi, R., Jacob, V. S., & Pirkul, H. (2004). The influence of communication mode and incentive structure on GDSS process and outcomes. *Decision Support Systems*, 37(2), 287–305.
https://doi.org/10.1016/S0167-9236(03)00023-X

Barley, S. R. (1991). Contextualizing conflict: Notes on the anthropology of disputes and negotiations. *Research on negotiation in organizations*, *3*, 165-199.

Bazerman, M. H., & Lewicki, R. (Eds.). (1983). *Negotiating in organizations* (Vol. 63). SAGE Publications, Incorporated.

Bazerman, M., Magliozzi, T., Neale, M. (1985). Integrative Bargaining in a Competitive Market. *Organizational Behavior and Human Decision Processes*, vol. 35(3), 294-313. https://doi.org/10.1016/0749-5978(85)90026-3

Bazerman, M. H., Curhan, J. R., Moore, D. A., & Valley, K. L. (2000). Negotiation. *Annual Review of Psychology*, 51(1), 279–314.
https://doi.org/10.1146/annurev.psych.51.1.279

Bazerman, M. H., & Moore, D. A. (2013). *Judgment in Managerial Decision Making*. Wiley. *New York*.

Bazerman, M. H., Neale, M. A., Valley, K. L., Zajac, E. J., & Kim, Y. M. (1992). The effect of agents and mediators on negotiation outcomes. *Organizational Behavior and Human Decision Processes*, 53(1), 55-73.
https://doi.org/10.1016/0749-5978(92)90054-B

Becker, T. E. (2005). Potential problems in the statistical control of variables in organizational research: A qualitative analysis with recommendations. *Organizational research methods*, *8*(3), 274-289.
https://doi.org/10.1177/1094428105278021

Bednar, D. A., & Curington, W. P. (1983). Interaction analysis: A tool for understanding negotiations. *ILR Review*, 36(3), 389-401. https://doi.org/10.1177/001979398303600305

Beersma, B., & De Dreu, C. K. (2002). Integrative and distributive negotiation in small groups: Effects of task structure, decision rule, and social motive. *Organizational Behavior and Human Decision Processes*, 87(2), 227-252. https://doi.org/10.1006/obhd.2001.2964

Ben-Shlomo, Y., Brookes, S., & Hickman, M. (2013). *Epidemiology, evidence-based medicine, and public health*. John Wiley & Sons.

Bercovitch, J., & Jackson, R. (2001a). Negotiation or mediation?: An exploration of factors affecting the choice of conflict management in international conflict. *Negotiation Journal*, *17*(1), 59-77. https://doi.org/10.1111/j.1571-9979.2001.tb00227.x

Bercovitch, J., & Jackson, R. (2001b). Current developments in international conflict management: Assessing the relevance of negotiation and mediation. *Cambridge Review of International Affairs*, *14*(2), 13-38. https://doi.org/10.1080/09557570108400354

Bergkvist, L., & Rossiter, J. R. (2007). The predictive validity of multiple-item versus single-item measures of the same constructs. *Journal of marketing research*, *44*(2), 175-184.

Bichler, M., Kersten, G., & Strecker, S. (2003). Towards a structured design of electronic negotiations. *Group Decision and Negotiation*, 12, 311-335. https://doi.org/10.1023/A:1024867820235

Bliss, C. (2000). Management von Komplexität: Ein Integrierter, Systemtheoretischer Ansatz zur Komplexitätsreduktion. *Management of Complexity: An Integrated, System-Theory Based Approach to Complexity Reduction*. Wiesbaden, Germany. Deutscher Universitäts-Verlag.

Blockus, M. O. (2010). *Komplexität in Dienstleistungsunternehmen: Komplexitätsformen, Kosten-und Nutzenwirkungen, empirische Befunde und Managementimplikationen* (Vol. 28). Springer-Verlag.

Bollen, K., & Euwema, M. (2013). Workplace Mediation: An underdeveloped research area. *Negotiation Journal*, 29(3), 329–353. https://doi.org/10.1111/nejo.12028

Bolton, A., Goosen, L., & Kritzinger, E. (2016, September). *Enterprise digitization enablement through unified communication & collaboration* [Paper presentation]. Proceedings of the Annual Conference of the South African Institute of Computer Scientists and Information Technologists, 1-10. https://doi.org/10.1145/2987491.2987516

Bothorel, G. (2020). The influence of brands' omni-channel strategy on consumer's behavior. *Business administration*. Université Panthéon-Sorbonne - Paris I. English. ffNNT: 2020PA01E031ff. fftel-03176560f

Braun, C., & Hadwich, K. (2016). Complexity of internal services: Scale development and validation. *Journal of Business Research*, *69*(9), 3508-3522. https://doi.org/10.1016/j.jbusres.2016.01.035

Breaugh, J. A. (2006). Rethinking the control of nuisance variables in theory testing. *Journal of Business and Psychology*, *20*, 429-443. https://doi.org/10.1007/s10869-005-9009-y

Breaugh, J. A. (2008). Important considerations in using statistical procedures to control for nuisance variables in non-experimental studies. *Human Resource Management Review*, *18*(4), 282-293. https://doi.org/10.1016/j.hrmr.2008.03.001

Brett, J. F., Pinkley, R. L., & Jackofsky, E. F. (1996). Alternatives to having a BATNA in dyadic negotiation: The influence of goals, self-efficacy, and alternatives on negotiated outcomes. *International Journal of Conflict Management*, *7*(2), 121-138. https://doi.org/10.1108/eb022778

Brett, J. M. (2014). *Negotiating Globally: How to Negotiate Deals, Resolve Disputes, and Make Decisions across Cultural Boundaries* (3rd ed.). Jossey-Bass.

Bretz. (1983). *Media for interactive communication*. SAGE Publications. http://www.gproxx.com/http://blogxd.info/dspace/uk/id-media-for-interactive-communication.pdf

Brown, J. L., & Agnew, N. M. (1982). Corporate agility. *Business Horizons*, 25(2), 29-33.

Brown, S., & Bessant, J. (2003). The manufacturing strategy-capabilities links in mass customisation and agile manufacturing – an exploratory study. *International Journal of Operations & Production Management*, *23*(7), 707-730. https://doi.org/10.1108/01443570310481522

Butler Jr, J. K. (1999). Trust expectations, information sharing, climate of trust, and negotiation effectiveness and efficiency. *Group & organization management*, 24(2), 217-238. https://doi.org/10.1177/1059601199242005

Burke, K., & Chidambaram, L. (1999). How much bandwidth is enough? A longitudinal examination of media characteristics and group outcomes. *MIS quarterly*, 557-579. https://doi.org/10.2307/249489

Cai, Z., Huang, Q., Liu, H., & Wang, X. (2018). Improving the agility of employees through enterprise social media: The mediating role of psychological conditions. *International Journal of Information Management*, *38*(1), 52-63. https://doi.org/10.1016/j.ijinfomgt.2017.09.001

Caputo, A., Fiorentino, R., & Garzella, S. (2019). From the boundaries of management to the management of boundaries: Business processes, capabilities, and negotiations. *Business Process Management Journal*, 25(3), 391-413. https://doi.org/10.1108/BPMJ-11-2017-0334

Carlson, K. D., & Wu, J. (2012). The illusion of statistical control: Control variable practice in management research. *Organizational research methods*, *15*(3), 413-435. https://doi.10.1177/1094428111428817

Carlson, J. R., & Zmud, R. W. (1999). Channel expansion theory and the experiential nature of media richness perceptions. *Academy of management journal*, *42*(2), 153-170.

Carnevale, P. J. D., Pruitt, D. G., & Seilheimer, S. D. (1981). Looking and competing: Accountability and visual access in integrative bargaining. *Journal of Personality and Social Psychology*, 40(1), 111–120. https://doi.org/10.1037/0022-3514.40.1.111

Carnevale, P. J. D., & Isen, A. M. (1986). The influence of positive affect and visual access on the discovery of integrative solutions in bilateral negotiation. *Organizational Behavior and Human Decision Processes*, 37(1), 1–13. https://doi.org/10.1016/0749-5978(86)90041-5

Carnevale, P. J., & Pruitt, D. G. (1992). Negotiation and Mediation. *Annual review of psychology*, 43(1), 531-582. https://doi.org/10.1146/annurev.ps.43.020192.002531

Chan, W. T. (2015). *The Way of Lao Tzu*. Ravenio Books.

Channa, N. A., Bhutto, M. H., Bhutto, M., Bhutto, N. A., & Tariq, B. (2022). Capturing customer's store loyalty through relationship benefits: moderating effect of retail innovation. *European Business Review*, 34(1), 20-40. https://doi.org/10.1108/EBR-09-2019-0179

Chang, Y., & Thorson, E. (2004). Television and web advertising synergies. *Journal of advertising*, *33*(2), 75-84. https://doi.org/10.1080/00913367.2004.10639161

Chalmers, I. (1993). The Cochrane collaboration: preparing, maintaining, and disseminating systematic reviews of the effects of health care. *Annals of the New York Academy of Sciences*, *703*, 156-63.

Charbonnier-Voirin, A. (2011). The development and partial testing of the psychometric properties of a measurement scale of organizational agility. *M@ n@ gement*, 119-156.

Chau, S. L. (2018). Enhancing Organizational Performance by Knowledge Sharing. *Journal of Business Theory and Practice*, Vol. 6, No. 2, 2018.

Chen, S., & Weiss, G. (2013). An efficient automated negotiation strategy for complex environments. *Engineering Applications of Artificial Intelligence*, 26(10), 2613–2623. https://doi.org/10.1016/j.engappai.2013.08.012

Child, J. (1972). Organizational Structure, Environment and Performance: The Role of Strategic Choice. *Sociology*, *6*(1), 1-22. https://doi.org/10.1177/003803857200600101

Chin, W. W., Marcolin, B. L., & Newsted, P. R. (1996) A partial least squares latent variable modeling approach for measuring interaction effects. Results from a Monte Carlo simulation study and voice mail emotion/adoption study. In Proceedings of the Seventeenth International Conference on Information Systems (DeGROSS, J. I., JARVENPAA, S., & SRINIVASAN, A., Eds), 21–41, Cleveland, OH.

Chin, W. W., Marcolin, B. L., & Newsted, P. R. (2003). A partial least squares latent variable modeling approach for measuring interaction effects: Results from a Monte Carlo simulation study and an electronic-mail emotion/adoption study. *Information systems research*, *14*(2), 189-217. https://doi.org/10.1287/isre.14.2.189.16018

Chiu, D. K. W., Cheung, S. C., Hung, P. C. K., Chiu, S. Y. Y., & Chung, A. K. K. (2005). Developing e-negotiation support with a meta-modeling approach in a web services environment. *Decision Support Systems*, 40(1), 51–69. https://doi.org/10.1016/j.dss.2004.04.004

Churchill Jr, G. A. (1979). A paradigm for developing better measures of marketing constructs. *Journal of marketing research*, *16*(1), 64-73. https://doi.org/10.1177/002224377901600110

Cilliers, P. (2000a). What can we learn from a theory of complexity?. *Emergence*, *2*(1), 23-33. https://doi.org/10.1207/S15327000EM0201_03

Cilliers, P. (2000b). Knowledge, complexity, and understanding. *Emergence, A Journal of Complexity Issues in Organizations and Management*, *2*(4), 7-13. https://doi.org/10.1207/S15327000EM0204_03

Cilliers, P. (2000c). Rules and complex systems. *Emergence, A Journal of Complexity Issues in Organizations and Management*, *2*(3), 40-50. https://doi.org/10.1207/S15327000EM0203_04

Citera, M., Beauregard, R., & Mitsuya, T. (2005). An experimental study of credibility in e-negotiations. *Psychology and Marketing*, 22(2), 163–179. https://doi.org/10.1002/mar.20053

Clark, H. H., & Wilkes-Gibbs, D. (1986). Referring as a collaborative process. *Cognition*, 22(1), 1-39. https://doi.org/10.1016/0010-0277(86)90010-7

Clark, H. H., & Brennan, S. E. (1991). Grounding in communication. In L. B. Resnick, J. M. Levine, & S. D. Teasley (Eds.), *Perspectives on socially shared cognition,* 127–149. American Psychological Association. https://doi.org/10.1037/10096-006

Clark, H. H., & Schaefer, E. F. (1989). Contributing to Discourse. *Cognitive Science*, 13(2), 259–294. https://doi.org/10.1207/s15516709cog1302_7

Cohen, J. (1988). *Statistical Power Analysis for the Behavioral Science* (2nd ed.). Hillsdale, NJ: Lawrence Erlbaum Associates.

Cohen, P., Cohen, J., Teresi, J., Marchi, M., & Velez, C. N. (1990). Problems in the measurement of latent variables in structural equations causal models. *Applied Psychological Measurement*, 14(2), 183-196.

Collier, J. E., & Sherrell, D. L. (2010). Examining the influence of control and convenience in a self-service setting. *Journal of the Academy of Marketing Science*, *38*, 490-509. https://doi.org/10.1007/s11747-009-0179-4

Coltman, T. R., Devinney, T. M., & Midgley, D. F. (2007). e-Business strategy and firm performance: a latent class assessment of the drivers and impediments to success. *Journal of Information Technology*, 22(2), 87-101. https://doi.10.1057/palgrave.jit.2000073

Connor-Smith, J. K., & Compas, B. E. (2002). Vulnerability to social stress: Coping as a mediator or moderator of sociotropy and symptoms of anxiety and depression. *Cognitive Therapy & Research*, 26(1). https://doi.org/10.1023/A:1013889504101

Cooke, R., & French, D. P. (2008). How well do the theory of reasoned action and theory of planned behaviour predict intentions and attendance at screening programmes? A meta-analysis. *Psychology and health*, *23*(7), 745-765. https://doi.org/10.1080/08870440701544437

Crook, T. R., & Combs, J. G. (2007). Sources and consequences of bargaining power in supply chains. *Journal of operations management*, 25(2), 546-555. https://doi.org/10.1016/j.jom.2006.05.008

Croson, R. T. A. (1999). Look at me when you say that: An electronic negotiation simulation. *Simulation & Gaming*, 30(1), 23–37. https://doi.org/10.1177/104687819903000105

Crump, L. (2011). Negotiation process and negotiation context. *International Negotiation*, 16(2), 197–227. https://doi.org/10.1163/138234011X573011

Crump, L., & Druckman, D. (2012). Turning points in multilateral trade negotiations on intellectual property. *International Negotiation*, 17(1), 9–35. https://doi.org/10.1163/157180612X630910

Cummins, S., Peltier, J. W., & Dixon, A. (2016). Omni-channel research framework in the context of personal selling and sales management: A review and research extensions. *Journal of Research in Interactive Marketing*, 10(1), 2–16. https://doi.org/10.1108/JRIM-12-2015-0094

Cunningham, J. B. (1977). Approaches to the evaluation of organizational effectiveness. *Academy of Management Review*, *2*(3), 463-474. https://doi.org/10.5465/amr.1977.4281841

Curtin, J., & Miller, R. (2011). Negotiating coalitions: Comparative perspectives. *Political Science*, 63(1), 3-9. https://doi.org/10.1177/0032318711403916

Daft, R. L., & Macintosh, N. B. (1981). A tentative exploration into the amount and equivocality of information processing in organizational work units. *Administrative science quarterly*, 207-224. https://doi.org/10.2307/2392469

Daft, R. L. & Lengel, R. H. (1984). Information richness: A new approach to managerial information processing and organization design, In L. L. Cummings & B. M. Staw (Eds.), Research in organizational behavior (Vol. 6, pp. 191-234). Greenwich, CT: JAI Press.

Daft, R. L., & Lengel, R. H. (1986). Organizational information requirements, media richness and structural design. *Management science*, *32*(5), 554-571. https://doi.org/10.1287/mnsc.32.5.554

Daft, R. L., Lengel, R. H., & Trevino, L. K. (1987). Message equivocality, media selection, and manager performance: Implications for information systems. *MIS quarterly*, 355-366. https://doi.org/10.2307/248682

Daft, R. L., & Lewin, A. Y. (1993). Where are the theories for the" new" organizational forms? An editorial essay. Organization science, i-vi. https://www.jstor.org/stable/2635077

Damian, D. (2002, May). The study of requirements engineering in global software development: as challenging as important. *Proceedings of Global Software Development, Workshop* (Vol. 9).

Damian, D. E., Eberlein, A., Shaw, M. L. G., & Gaines, B. R. (2003). An exploratory study of facilitation in distributed requirements engineering. *Requirements Engineering*, 8(1), 23–41. https://doi.org/10.1007/s00766-002-0164-7

Damian, D., Lanubile, F., & Mallardo, T. (2008). On the need for mixed media in distributed requirements negotiations. *IEEE Transactions on Software Engineering*, 34(1), 116–132. https://doi.org/10.1109/TSE.2007.70758

Danaher, P., & Rossiter, J. R. (2006). A comparison of the effectiveness of marketing communication channels: Perspectives from both receivers and senders. *Occasional Papers University of Auckland, New Zealand.*

Danaher, P., & Rossiter, J. (2011). Comparing perceptions of marketing communication channels. *European Journal of Marketing*, 45, 6-42. https://doi.10.1108/03090561111095586.

Dalkey, N. C., Brown, B. B., & Cochran, S. (1969). *The Delphi method: An experimental study of group opinion* (Vol. 3, p. 107). Santa Monica, CA: Rand Corporation.

Dardas, L. A., & Ahmad, M. M. (2015). Coping strategies as mediators and moderators between stress and quality of life among parents of children with autistic disorder. *Stress and Health*, 31(1), 5-12. https://doi.org/10.1002/smi.2513

Dawson, J. F. (2014). Moderation in management research: What, why, when, and how. *Journal of Business and Psychology*, *29*(1), 1-19. https://doi.org/10.1007/s10869-013-9308-7

De Battisti, F., & Siletti, E. (2019). On the use of control variables in PLS-SEM= Sull'Uso delle Variabili di Controllo nei PLS-SEM. In *Smart Statistics for Smart Applications: Book of Short Papers* (pp. 793-798). Pearson.

De Dreu, C. K., & Van Kleef, G. A. (2004a). The influence of power and parties on negotiation. *Negotiation theory and research*, 255-276. Psychology Press.

De Dreu, C. K., & Van Kleef, G. A. (2004b). The influence of power on the information search, impression formation, and demands in negotiation. *Journal of Experimental Social Psychology*, *40*(3), 303-319. https://doi.org/10.1016/j.jesp.2003.07.004

Delaney, M. M., Foroughi, A., & Perkins, W. C. (1997). An empirical study of the efficacy of a computerized negotiation support system (NSS). *Decision Support Systems*, 20(3), 185–197. https://doi.org/10.1016/S0167-9236(96)00051-6

Delbecq, A. L., & Van de Ven, A. H. (1971). A group process model for problem identification and program planning. *The journal of applied behavioral science*, *7*(4), 466-492. https://doi.org/10.1177/002188637100700404

Dennis, A. R., & Kinney, S. T. (1998). Testing media richness theory in the new media: The effects of cues, feedback, and task equivocality. *Information systems research*, 9(3), 256-274. https://doi.org/10.1287/isre.9.3.256

Dennis, A. R., Fuller, R. M., & Valacich, J. S. (2008). Media, tasks, and communication processes: A theory of media synchronicity. *MIS quarterly*, 575-600. https://doi.org/10.2307/25148857

Dennis, A. R., & Robert, M. F. (2009). Media synchronicity and media choice: Choosing media for performance. *Media Choice*, 261-287. Routledge.

De Meuse, K. P., Dai, G., & Hallenbeck, G. S. (2010). Learning agility: A construct whose time has come. *Consulting Psychology Journal: Practice and Research*, *62*(2), 119. https://doi.org/10.1037/a0019988

De Paula, M. V., de Souza, J. M., Blaschek, J. R., & Baião, F. (2004, January). *A cooperative system to support inventory leveling negotiations* [Paper presentation]. 37th Annual Hawaii International Conference on System Sciences, 2004. Proceedings of the, 10. IEEE. https://doi.10.1109/HICSS.2004.1265457

DeSanctis, G., & Poole, M. S. (1994). Capturing the complexity in advanced technology use: Adaptive structuration theory. *Organization science*, *5*(2), 121-147. https://doi.org/10.1287/orsc.5.2.121

Diamond, J., & Weiss, R. (2013). *Negotiation: Theory and practice*. Harvard Business Press.

Dijkstra, M., Buijtels, H. E., & Van Raaij, W. F. (2005). Separate and joint effects of medium type on consumer responses: a comparison of television, print, and the Internet. *Journal of Business Research*, *58*(3), 377-386. https://doi.org/10.1016/S0148-2963(03)00105-X

Dijkstra, T.K. (2010). Latent Variables and Indices: Herman Wold's Basic Design and Partial Least Squares. In: Esposito Vinzi, V., Chin, W., Henseler, J., Wang, H. (eds) Handbook of Partial Least Squares. *Springer Handbooks of Computational Statistics*. Springer, Berlin, Heidelberg. https://doi.org/10.1007/978-3-540-32827-8_2

Dijkstra, T. K. (2014). PLS'Janus face–response to professor Rigdon's 'rethinking partial least squares modeling: in praise of simple methods'. *Long Range Planning*, *47*(3), 146-153. https://doi.org/10.1016/j.lrp.2014.02.004

Dijkstra, T. K., & Henseler, J. (2015). Consistent Partial Least Squares Path Modeling. *MIS Quarterly*, *39*(2), 297–316. https://www.jstor.org/stable/26628355

Dixon, T. M., & Baumeister, R. F. (1991). Escaping the self: The moderating effect of self-complexity. *Personality and Social Psychology Bulletin*, 17(4), 363-368. https://doi.org/10.1177/0146167291174002

Donohue, W. A., Ramesh, C., & Borchgrevink, C. (1991). Crisis bargaining: Tracking relational paradox in hostage negotiation. *International Journal of Conflict Management*, 2(4), 258-274. https://doi.org/10.1108/eb022701

Donohue, W. A., & Roberto, A. J. (1996). An empirical examination of three models of integrative and distributive bargaining. *International Journal of Conflict Management*, *7*(3), 209-229. https://doi.org/10.1108/eb022782

Dorado, M. A., Medina, F. J., Munduate, L., Cisneros, I. F. J., & Euwema, M. (2002). Computer-mediated negotiation of an escalated conflict. *Small Group Research*, 33(5), 509–524. https://doi.org/10.1177/104649602237168

Dorothea Brack, A., & Benkenstein, M. (2014). Responses to other similar customers in a service setting – analyzing the moderating role of perceived performance risk. *Journal of Services Marketing*, 28(2), 138-146. https://doi.org/10.1108/JSM-05-2012-0089

Dove, R. (2001). *Response Ability: The Language, Structure, and Culture of the Agile Enterprise*. New York: Wiley

Drolet, A. L., & Morrison, D. G. (2001). Do we really need multiple-item measures in service research? *Journal of service research*, *3*(3), 196-204. https://doi.org/10.1177/109467050133001

Drucker, P. (1974). *Management* (1st ed.). Routledge. https://doi.org/10.4324/9780080939063

Dupont, C., & Faure, G. O. (1991). The negotiation process. *International negotiation*, 40-57.

Dwyer, S. (2018). *Successful negotiation: Essential strategies and skills*. Routledge.

Dyer, J. H., Singh, H., & Hesterly, W. S. (2018). The relational view revisited: A dynamic perspective on value creation and value capture. *Strategic management journal*, 39(12), 3140-3162. https://doi.org/10.1002/smj.2785

Ebner, N. (2017). Negotiation is changing. *Journal of Dispute Resolution*, 99.

Edell, J. A., & Keller, K. L. (1989). The information processing of coordinated media campaigns. *Journal of Marketing Research*, *26*(2), 149-163. https://doi.org/10.1177/002224378902600202

Edström, A., & Galbraith, J. R. (1977). Transfer of managers as a coordination and control strategy in multinational organizations. *Administrative science quarterly*, 248-263. https://doi.org/10.2307/2391959

Edwards, J. R., & Bagozzi, R. P. (2000). On the nature and direction of relationships between constructs and measures. *Psychological methods*, *5*(2), 155.

Eggert, A., Fassott, G., & Helm, S. (2005). Identifizierung und Quantifizierung mediierender und moderierender Effekte in komplexen Kausalstrukturen. *Handbuch PLS-Pfadmodellierung, Stuttgart*, 101-116.

Eidhoff, A. T. (2017). Digitalization and the industrial sales process: A configurational analysis of digital and non-digital sales phases. *Hohenheimer Arbeits- und Projektberichte zum Marketing* Nr. 23.

Eklinder-Frick, J. O., & Åge, L. J. (2020). Relational business negotiation - propositions based on an interactional perspective. Journal of Business & Industrial Marketing, 35(5), 925-937. https://doi.org/10.1108/JBIM-04-2019-0169

El-Shinnawy, M. M., & Markus, M. L. (1992). Media richness theory and new electronic communication media: A study of voice mail and electronic mail.

ElShenawy, E. (2010) Does Negotiation Training improve Negotiator's Performance? *Journal of European Industrial Training*, vol. 34 (3), 192-210. https://doi.org/10.1108/03090591011031719

Ertel, D. (1999). Turning negotiation into a corporate capability. *Harvard Business Review*, 77(3), 55-56. https://link.gale.com/apps/doc/A54556298/AONE?u=anon~72642e6b&sid=googleScholar&xid=efa652ff

Es, R. van, French, W., & Stellmaszek, F. (2004). Resolving conflicts over Ethical issues: Face-to-face versus internet negotiations. *Journal of Business Ethics*, 53(1/2), 165–172. https://doi.org/10.1023/B:BUSI.0000039406.58170.67

Espinosa, J. A., Nan, N., & Carmel, E. (2015). Temporal distance, communication patterns, and task performance in teams. *Journal of Management Information Systems*, *32*(1), 151-191. https://doi.org/10.1080/07421222.2015.1029390

Evans, D. (2004, October). An introduction to unified communications: challenges and opportunities. *Aslib Proceedings,* Vol. 56, No. 5, 308-314. Emerald Group Publishing Limited. https://doi.org/10.1108/00012530410560904

Everitt, B., Landau, S., & Leese, M. (2001). *Cluster Analysis.* A Hodder Arnold Publication. Willey: London, UK.

Falk, R. F., & Miller, N. B. (1992). *A primer for soft modeling*. University of Akron Press.

Faizullaev, A. (2014). Diplomatic interactions and negotiations. *Negotiation journal*, 30(3), 275-299. https://doi.org/10.1111/nejo.12061

Fang, T. (2006). Negotiation: The Chinese style. *Journal of Business & Industrial Marketing*. https://doi.org/10.1108/08858620610643175

Fassott, G., & Eggert, A. (2005). Zur Verwendung formativer und reflektiver Indikatoren in Strukturgleichungsmodellen: Bestandsaufnahme und Anwendungsempfehlungen. *Handbuch PLS-Pfadmodellierung: Methode, Anwendung, Praxisbeispiele, Stuttgart, 33.*

Fells, R., Rogers, H., Prowse, P., & Ott, U. F. (2015). Unraveling business negotiations using practitioner data. *Negotiation and Conflict Management Research*, 8(2), 119-136. https://doi.org/10.1111/ncmr.12050

Fernandes, C., Farinha, J., Martins, F. V., & Mateus, C. (2018). Bank governance and performance: A survey of the literature. *Journal of Banking Regulation*, 19, 236-256. https://doi.org/10.1057/s41261-017-0045-0

Fish, R. S., Kraut, R. E., Root, R. W. & Rice, R E (1993). Video as a Technology for Informal Communication. Communications of the ACM, January 1993, Vol. 36, No 1, 48- 61.

Fishbein, M., & Ajzen, I. (2010). *Predicting and changing behavior: The reasoned action approach*. Taylor & Francis.

Fisher, R., Ury, W. L., & Patton, B. (2011). *Getting to yes: Negotiating agreement without giving in*. Penguin.

Fisher, J., & Fisher-Yoshida, B. (2017). Educating negotiators: Using theory, practice, case studies, and simulations in an integrated learning experience. *Negotiation and Conflict Management Research*, 10(4), 286-305. https://doi.org/10.1111/ncmr.12104

Flanagin, A. J., & Metzger, M. J. (2001). Internet use in the contemporary media environment. *Human communication research*, *27*(1), 153-181. https://doi.org/10.1111/j.1468-2958.2001.tb00779.x

Forman, J., & Damschroder, L. (2007). Qualitative content analysis. In *Empirical methods for bioethics: A primer*, Vol. 11, 39-62. Emerald Group Publishing Limited.

Fornell, C., & Larcker, D. F. (1981). Structural equation models with unobservable variables and measurement error: *Algebra and statistics*. https://doi.org/10.1177/002224378101800313

Fry, L. W. (1982). Technology-structure research: Three critical issues. *Academy of Management Journal*, 25(3), 532-552. https://doi.org/10.5465/256026

Fry, W. R. (1985). The effect of dyad Machiavellianism and visual access on integrative bargaining outcomes. *Personality and Social Psychology Bulletin*, 11(1), 51–62. https://doi.org/10.1177/0146167285111005

Fulk, J., Schmitz, J., & Steinfield, C. W. (1990). A social influence model of technology use. *Organizations and communication technology*, 117, 140.

Galbraith, J. (1973). *Designing complex organizations*. Reading, Mass.

Galin, A., Gross, M., & Gosalker, G. (2007). E-negotiation versus face-to-face negotiation what has changed – if anything?. *Computers in Human Behavior*, *23*(1), 787-797. https://doi.org/10.1016/j.chb.2004.11.009

Galinsky, A., & Schweitzer, M. (2015). *Friend & foe: When to cooperate, when to compete, and how to succeed at both*. Crown Business. New York.

Ganesan, S. (1993). Negotiation strategies and the nature of channel relationships. *Journal of marketing research*, 30(2), 183-203. https://doi.org/10.1177/002224379303000205

Gaines, B. R., Shaw, M. L. G., Eberlein, A., & Damian, D. E. H. (2000). Using different communication media in requirements negotiation. *IEEE Software*, 17(3), 28–36. https://doi.org/10.1109/52.896247

Geiger, I. (2014). Media effects on the formation of negotiator satisfaction: The example of face-to-face and text-based electronically mediated negotiations. *Group Decision and Negotiation*, 23(4), 735–763. https://doi.org/10.1007/s10726-012-9317-3

Geiger, I. (2020). From letter to twitter: A systematic review of communication media in negotiation. *Group Decision and Negotiation*, 29(2), 207–250. https://doi.org/10.1007/s10726-020-09662-6

Geiger, I. (2021). "From letter to Twitter: A systematic review of communication media in negotiation": Correction. *Group Decision and Negotiation, 30*(5), 1193–1194. https://doi.org/10.1007/s10726-021-09751-0

Geiger, I., & Parlamis, J. (2014). Is there more to email negotiation than email? The role of email affinity. *Computers in Human Behavior*, 32, 67–78. https://doi.org/10.1016/j.chb.2013.11.016

Geiger, I., & Laubert, C. (2018). Situational strategic versus personal influences on negotiation medium choice: Media synchronicity theory and affect for communication channel. *International Journal of Conflict Management*, 29(3), 398–423. https://doi.org/10.1108/IJCMA-06-2017-0054

Geiger, I., & Hüffmeier, J. (2020). “The more, the merrier” or “less is more”? How the number of issues addressed in B2B sales negotiations affects dyadic and seller economic outcomes. *Industrial Marketing Management*, 87, 90-105. https://doi.org/10.1016/j.indmarman.2020.02.013

Gelfand, M. J., & Dyer, L. (Eds.). (2010). *The Oxford handbook of negotiation*. Oxford University Press.

Ghauri, P. N., & Usunier, J. C. (Eds.). (2003). *International business negotiations*. Emerald Group Publishing.

Gist, M. E., Stevens, C. K., & Bavetta, A. G. (1991). Effects of self-efficacy and post-training intervention on the acquisition and maintenance of complex interpersonal skills. *Personnel psychology*, *44*(4), 837-861. https://doi.org/10.1111/j.1744-6570.1991.tb00701.x

Gorsuch, R. L. (1997). Exploratory factor analysis: Its role in item analysis. *Journal of personality assessment*, 68(3), 532-560. https://doi.org/10.1207/s15327752jpa6803_5

Graham, J. L. (1986). The problem-solving approach to negotiations in industrial marketing. *Journal of Business Research*, 14(6), 549-566. https://doi.org/10.1016/0148-2963(86)90014-7

Graham, J. L. (1987). A theory of interorganizational negotiations. *Research in marketing*.

Graham, J. L., Mintu, A. T., & Rodgers, W. (1994). Explorations of negotiation behaviors in ten foreign cultures using a model developed in the United States. *Management Science*, 40(1), 72-95. https://doi.org/10.1287/mnsc.40.1.72

Grennan, M. (2014). Bargaining ability and competitive advantage: Empirical evidence from medical devices. *Management Science*, 60(12), 3011-3025. https://doi.org/10.1287/mnsc.2014.2006

Greenhalgh, L., & Chapman, D. I. (1995). Joint decision making. The inseperability of relationships and negotiation. *Negotiation as a social process*, 166-185. SAGE Publications.

Griffith, T. L., & Northcraft, G. B. (1994). Distinguishing between the forest and the trees: Media, features, and methodology in electronic communication research. *Organization Science*, *5*(2), 272-285. https://doi.org/10.1287/orsc.5.2.272

Groves, K. S., Feyerherm, A., & Gu, M. (2015). Examining cultural intelligence and cross-cultural negotiation effectiveness. *Journal of Management Education*, *39*(2), 209-243. https://doi.org/10.1177/1052562914543273

Gulati, R., & Garino, J. (2000). Get the Right Mix of Bricks & Clicks. *Harvard Business Review*, 78(3), 107. https://link.gale.com/apps/doc/A63035682/AONE?u=anon~724848d3&sid=googleScholar&xid=e386d4af

Haggenmüller, S., Oehlschläger, P., Herbst, U., & Voeth, M. (2023). Time for change? Scenario analysis on buyer–seller negotiations. *Journal of Business & Industrial Marketing*, 38(5), 1215-1242. https://doi.org/10.1108/JBIM-11-2021-0511

Hair, J. F., Black, W. C., Babin, B. J., & Anderson, R. E. (2010). *Multivariate data analysis*: Global edition: Pearson Higher Education Upper Saddle River.

Hair, J. F., Ringle, C. M., & Sarstedt, M. (2013). Partial least squares structural equation modeling: Rigorous applications, better results, and higher acceptance. *Long range planning*, *46*(1-2), 1-12. https://ssrn.com/abstract=2233795

Hair, J. F., Black, W. C., Babin, B. J., & Anderson, R. E. (2014). *Multivariate data analysis* (7th ed.). Upper Saddle River, NJ: Pearson Education.

Hair Jr, J. F., Sarstedt, M., Matthews, L. M., & Ringle, C. M. (2016). Identifying and treating unobserved heterogeneity with FIMIX-PLS: part I–method. *European business review*, *28*(1), 63-76. https://doi.org/10.1108/EBR-09-2015-0094

Hair Jr, J. F., Hult, G. T. M., Ringle, C., & Sarstedt, M. (2017). *A primer on partial least squares structural equation modeling (PLS-SEM)*. Sage Publications.

Hair, J. F., Risher, J. J., Sarstedt, M., & Ringle, C. M. (2019). When to use and how to report the results of PLS-SEM. *European Business Review*, 31(1), 2-24. https://doi.org/10.1108/EBR-11-2018-0203

Hair Jr, J. F., Hult, G. T. M., Ringle, C. M., Sarstedt, M., Danks, N. P., & Ray, S. (2021). *Partial least squares structural equation modeling (PLS-SEM) using R: A workbook* (p. 197). Springer Nature. https://doi.org/10.1007/978-3-319-57413-4_15

Hair, J. F., Hult, G. T. M., Ringle, C. M., & Sarstedt, M. (2022). *A primer on partial least squares structural equation modeling (PLS-SEM)* (3 ed.). Thousand Oaks, CA: Sage.

Halevy, N., & Phillips, L. T. (2015). Conflict templates in negotiations, disputes, joint decisions, and tournaments. *Social Psychological and Personality Science*, *6*(1), 13-22.

Harinck, F., & De Dreu, C. K. (2008). Take a break! or not? The impact of mindsets during breaks on negotiation processes and outcomes. *Journal of Experimental Social Psychology*, 44(2), 397-404. https://doi.org/10.1016/j.jesp.2006.12.009

Harris, K. L. (1996). Content analysis in negotiation research: A review and guide. *Behavior Research Methods, Instruments, & Computers*, *28*(3), 458-467. https://doi.org/10.3758/BF03200525

Hartmann, T. (2009). A brief introduction to media choice. *Media choice. A theoretical and empirical overview*, 1-9.

Hartmann, T. (2010, September). *Exploring the influence of media richness on engineering task performance and user perceptions* [Paper presentation]. *Proceedings of the european conference on product and process modelling*, 243-249.

Haselhuhn, M. P., Wong, E. M., Ormiston, M. E., Inesi, M. E., & Galinsky, A. D. (2014). Negotiating face-to-face: Men's facial structure predicts negotiation performance. *The Leadership Quarterly*, 25(5), 835–845. https://doi.org/10.1016/j.leaqua.2013.12.003

Hayduk, L. A. (1987). *Structural equation modeling with LISREL: Essentials and advances*. Jhu Press.

Hayles, N. K. (2012). *How we think: Digital media and contemporary technogenesis*. University of Chicago Press.

Heeter, C., Brown, N., Soffin, S., Stanley, C., & Salwen, M. (1989). Agenda-setting by electronic text news. *Journalism Quarterly*, 66(1), 101–106. https://doi.org/10.1177/107769908906600113

Heinen, J. S., & Jacobson, E. (1976). A model of task group development in complex organizations and a strategy of implementation. *Academy of Management Review*, *1*(4), 98-111. https://doi.org/10.5465/amr.1976.4396620

Hempel, C. G., & Oppenheim, P. (1948). Studies in the Logic of Explanation. *Philosophy of science*, *15*(2), 135-175.

Henderson, M. D. (2011). Mere physical distance and integrative agreements: When more space improves negotiation outcomes. *Journal of Experimental Social Psychology*, 47(1), 7-15. https://doi.org/10.1016/j.jesp.2010.07.011

Henseler, J., & Chin, W. W. (2010). A comparison of approaches for the analysis of interaction effects between latent variables using partial least squares path modeling. *Structural Equation Modeling*, *17*(1), 82-109. https://doi.org/10.1080/10705510903439003

Henseler, J., Ringle, C. M., & Sarstedt, M. (2015). A new criterion for assessing discriminant validity in variance-based structural equation modeling. *Journal of the academy of marketing science*, *43*, 115-135. https://doi.org/10.1007/s11747-014-0403-8

Herbst, U., & Merz, M. A. (2011). The industrial brand personality scale: Building strong business-to-business brands. *Industrial marketing management*, 40(7), 1072-1081. https://doi.org/10.1016/j.indmarman.2011.09.003

Herbst, U., Voeth, M., & Meister, C. (2011). What do we know about buyer–seller negotiations in marketing research? A status quo analysis. *Industrial Marketing Management*, 40(6), 967-978. https://doi.org/10.1016/j.indmarman.2011.07.004

Herbst, U., & Schwarz, S. (2011). How valid is negotiation research based on student sample groups? New insights into a long-standing controversy. *Negotiation Journal*, 27(2), 147-170. https://doi.org/10.1111/j.1571-9979.2011.00300.x

Herbst, U., Kemmerling, B., & Neale, M. A. (2017). All in, one-at-a-time or somewhere in the middle? Leveraging the composition and size of the negotiating package. *Journal of Business & Industrial Marketing*, 32(4), 580-586. https://doi.org/10.1108/JBIM-12-2015-0251

Herbst, U., Voeth. M., Hefner, M., Siebert, E. (2019) *Gain without Pain: How to Make Optimal Negotiation Packages* [Paper presentation]. Presentation at the 32nd International Association for Conflict Management (IACM 2019). Dublin.

Herrmann, A., Huber, F., & Kressmann, F. (2006). Varianz-und kovarianzbasierte Strukturgleichungsmodelle — ein Leitfaden zu deren Spezifikation, Schätzung und Beurteilung. *Schmalenbachs Zeitschrift für betriebswirtschaftliche Forschung*, *58*, 34-66. https://doi.org/10.1007/BF03371643

Hildebrandt, L., & Homburg, B. C. (1998). Die Kausalanalyse: Bestandsaufnahme, Entwicklungsrichtungen, Problemfelder in: Die Kausalanalyse, L. Hildenbrandt/C. Homburg (Publishers), Stuttgart 1998, p. 20.

Hollander-Blumoff, R. (2017). Formation of procedural justice judgments in legal negotiation. *Group Decision and Negotiation*, 26, 19-43. https://doi.org/10.1007/s10726-016-9498-2

Hollingshead, A. B., McGrath, J. E., & O'Connor, K. M. (1993). Group task performance and communication technology: A longitudinal study of computer-mediated versus face-to-face work groups. *Small Group Research*, 24(3), 307–333. https://doi.org/10.1177/1046496493243003

Holmlund-Rytkönen, M., & Strandvik, T. (2005). Stress in business relationships. *Journal of Business & Industrial Marketing*, 20(1), 12-22. https://doi.org/10.1108/08858620510576757

Homburg, C. & Dobratz, A. (1992). Covariance structure analysis via specification searches. Statistical Papers 33 (1), 119-142.

Homburg, C. (1996). Giering, A.(1996): Konzeptualisierung und Operationalisierung komplexer Konstrukte. *Marketing ZFP*, *18*(1), 5-24.

Hopp, W. J., & Oyen, M. P. (2004). Agile workforce evaluation: a framework for cross-training and coordination. *Iie Transactions*, *36*(10), 919-940. https://doi.org/10.1080/07408170490487759

Hosein, Z., & Yousefi, A. (2012). The role of emotional intelligence on workforce agility in the workplace. *International Journal of Psychological Studies*, *4*(3), 1-48.

Huang, X., Gattiker, T. F., & Schwarz, J. L. (2008). Interpersonal trust formation during the supplier selection process: The role of the communication channel. *Journal of Supply Chain Management*, 44(3), 53–75. https://doi.org/10.1111/j.1745-493X.2008.00066.x

Huber, F., Herrmann, A., Meyer, F., Vogel, J., Vollhardt, K. (2007). Kausalmodellierung mit Partial Least Squares – Anwendungsorientierte Einführung. Gabler Verlag, Wiesbaden. https://doi:10.1007/978-3-8349-9328-1

Hyder, E. B., Prietula, M. J., & Weingart, L. R. (2000). Getting to best: Efficiency versus optimality in negotiation. *Cognitive Science*, *24*(2), 169-204. https://doi.org/10.1207/s15516709cog2402_1

Iklé, F. C. *How Nations Negotiate.* New York: Harper and Row, 1964.

Jaccard, J., & Jacoby, J. (2010). *Theory construction and model-building skills: A practical guide for social scientists.* Guilford Press.

Jaccard, J., & Wan, C. K. (1996). *LISREL approaches to interaction effects in multiple regression* (No. 114). sage.

Jarvis, C. B., MacKenzie, S. B., & Podsakoff, P. M. (2003). A critical review of construct indicators and measurement model misspecification in marketing and consumer research. *Journal of consumer research*, *30*(2), 199-218. https://doi.org/10.1086/376806

Jennings, N. R., Parsons, S., Norriega, P., & Sierra, C. (1998). *On augmentation-based negotiation.*

Jiang, H., Zhou, D., & Zhang, J. H. (2019). Analysts' information acquisition and stock price synchronicity: A regulatory perspective from China. *Accounting Horizons*, 33(1), 153-179. https://doi.org/10.2308/acch-52259

Jöreskog, K. G. (1971). Statistical analysis of sets of congeneric tests. *Psychometrika*, *36*(2), 109-133. https://doi.org/10.1007/BF02291393

Johnson, N. A., & Cooper, R. B. (2015). Understanding the influence of instant messaging on ending concessions during negotiations. *Journal of Management Information Systems*, 31(4), 311-342. https://doi.org/10.1080/07421222.2014.1001285

Johnston, M. W., & Marshall, G. W. (2016). *Sales force management: Leadership, innovation, technology.* Routledge.

Judge, T. A., Erez, A., Bono, J. E., & Thoresen, C. J. (2002). Are measures of self-esteem, neuroticism, locus of control, and generalized self-efficacy indicators of a common core construct?. *Journal of personality and social psychology*, *83*(3), 693. https://doi.org/10.1037/0022-3514.83.3.693

Jung, J. Y., & Moro, M. (2014). Multi-level functionality of social media in the aftermath of the Great East Japan Earthquake. *Disasters*, 38(2), 123-143. https://doi.org/10.1111/disa.12071

Kachelmeier, S. J., & Towry, K. L. (2002). Negotiated transfer pricing: Is fairness easier said than done?. *The Accounting Review*, *77*(3), 571-593. https://doi.org/10.2308/accr.2002.77.3.571

Kaspzark, S. (2017). *Verhandlungskosten: Relevanz, Einflussfaktoren und Wirkung* (Doctoral Dissertation). Cumulative thesis in Negotiation Management. Schriftenreihe zum Verhandlungsmanagement. Publisher Verlag Dr. Kovac

Katz, E., Blumler, J. G., & Gurevitch, M. (1973). Uses and gratifications research. *The public opinion quarterly*, 37(4), 509-523.

Katz, D., & Kahn, R. L. (1978a). Organisation als System. *Handlungssysteme*, 217-231. https://doi.10.1007/978-3-322-85709-5_15

Katz, D., & Kahn, R. L. (1978b). *The social psychology of organizations*, Vol. 2, 528. New York: wiley.

Katz, D., & Kahn, R. L. (1978c). Organizations and the system concept. *Classics of organization theory*, *80*, 480.

Keh, H. T., & Pang, J. (2010). Customer reactions to service separation. *Journal of Marketing*, 74(2), 55-70.

Kenny, D. A. (2016). *Moderation*. http://davidakenny.net/cm/moderation.htm

Kenny, D. A. (2018). Moderator variables: introduction. *Moderator Variables: Introduction*.

Kersten, G. E., & Lai, H. (2007). Negotiation support and e-negotiation systems: An overview. *Group Decision and Negotiation*, 16, 553-586. https://doi.org/10.1007/s10726-007-9095-5

Kessler, F. (2008). *Instant Messaging. Eine neue interpersonale Kommunikationsform*. Networx Nr. 52, 52.

Khan, H. H., Malik, N., Usman, M., & Ikram, N. (2011). *Impact of Changing Communication Media on Conflict Resolution in Distributed Software Development Projects*, 189–194. https://doi.org/10.1109/MySEC.2011.6140667

Kidd, P.T. (Ed.) (*1994*), *A 21st Century Paradigm. In Agile Manufacturing: Forging New Frontiers*, Addison-Wesley, Wokingham.

Kiely, R. (2016). *The rise and fall of emerging powers: Globalisation, US Power and the Global North-South divide*. Springer.

Kiesler, S., Siegel, J., & McGuire, T. W. (1984). Social psychological aspects of computer-mediated communication. *American psychologist*, *39*(10), 1123. https://doi.org/10.1037/0003-066X.39.10.1123

Kiesler, S., & Sproull, L. (1992). Group decision making and communication technology. *Organizational Behavior and Human Decision Processes*, 52(1), 96–123. https://doi.org/10.1016/0749-5978(92)90047-B

Kietzmann, J. H., Hermkens, K., McCarthy, I. P., & Silvestre, B. S. (2011). Social media? Get serious! Understanding the functional building blocks of social media. *Business horizons*, 54(3), 241-251. https://doi.org/10.1016/j.bushor.2011.01.005

King, & Xia. (1997). Media appropriateness: effects of experience on communication media choice. *Decision Sciences*, 28(4), 877–910. https://doi.org/10.1111/j.1540-5915.1997.tb01335.x

Kitchen, P. J., & Schultz, D. E. (1999, January). A Multi-Country Comparison of the Drive for IMC. *Journal of Advertising Research*, *39*(1), 21. https://link.gale.com/apps/doc/A63669925/AONE?u=anon~4df0aa9d&sid=googleScholar&xid=e61c00ee

Kirk, A. K., & Brown, D. F. (2003). Latent constructs of proximal and distal motivation predicting performance under maximum test conditions. *Journal of Applied Psychology*, *88*(1), 40. https://doi.org/10.1037/0021-9010.88.1.40

Knöpfle, T. A. (2018). *Methoden der Datenerhebung in der Verhandlungsforschung: Erkenntnisse und Trends einer literaturbasierten Analyse und Darlegung zukünftiger Perspektiven* (Doctoral Dissertation). Cumulative thesis in Negotiation Management. Publisher Dr. Kovac.

Koch, E. J., & Shepperd, J. A. (2004). Is self-complexity linked to better coping? A review of the literature. *Journal of personality*, 72(4), 727-760. https://doi.org/10.1111/j.0022-3506.2004.00278.x

Kock, N. (2005). Media richness or media naturalness? The evolution of our biological communication apparatus and its influence on our behavior toward e-communication tools. *IEEE transactions on professional communication*, *48*(2), 117-130. https://doi.10.1109/TPC.2005.849649

Kohlbacher, F. (2006). The use of qualitative content analysis in case study research. *Forum Qualitative Sozialforschung/Forum: Qualitative Social Research*, Vol. 7, No. 1, 1-30. Institut fur Klinische Sychologie and Gemeindesychologie. https://doi.10.17169/fqs-7.1.75

Kulik, C. T., & Olekalns, M. (2012). Negotiating the gender divide: Lessons from the negotiation and organizational behavior literatures. *Journal of Management*, 38(4), 1387-1415. https://doi.10.1177/0149206311431307

Kunz, R., & Kunz, R. (2014a). Quantitative Hauptstudie: Mobile TV-Nutzung und Sportrezeption. Sportinteresse und Mobile TV. *Eine empirische Analyse der Einflussfaktoren des Nutzungsverhaltens*, 104-224. https://doi.org/10.1007/978-3-8349-4515-0_4

Kunz, R., & Kunz, R. (2014b). Darlegung der theoretischen Basis: Konsum neuer Sportmedien. Sportinteresse und Mobile TV. *Eine empirische Analyse der Einflussfaktoren des Nutzungsverhaltens*, 9-86. https://doi.org/10.1007/978-3-8349-4515-0_2

Kumar, A., & Motwani, J. (1995). A methodology for assessing time-based competitive advantage of manufacturing firms. *International Journal of Operations & Production Management, 15*(2), 36-53. https://doi.org/10.1108/01443579510080409

Kyl-Heku, L. M., & Buss, D. M. (1996). Tactics as units of analysis in personality psychology: An illustration using tactics of hierarchy negotiation. *Personality and Individual Differences*, 21(4), 497-517. https://doi.org/10.1016/0191-8869(96)00103-1

Lasswell, H. D. (1948). The structure and function of communication in society. In L. Bryson (Ed.), The communication of ideas (pp. 37-51). New York: Harper and Row.

Latané, B. (1981). The psychology of social impact. *American Psychologist*, 36(4), 343-356. https://doi.org/10.1037/0003-066X.36.4.343

Latif, K. F., Sajjad, A., Bashir, R., Shaukat, M. B., Khan, M. B., & Sahibzada, U. F. (2020). Revisiting the relationship between corporate social responsibility and organizational performance: The mediating role of team outcomes. *Corporate Social Responsibility and Environmental Management*, 27(4), 1630-1641. https://doi.org/10.1002/csr.1911

Lavie, D. (2007). Alliance portfolios and firm performance: A study of value creation and appropriation in the US software industry. *Strategic management journal*, 28(12), 1187-1212. https://doi.org/10.1002/smj.637

Lax, D. A., & Sebenius, J. K. (1986). Interests: The measure of negotiation. *Negotiation Journal*, *2*(1), 73-92. https://doi.org/10.1007/BF00998936

Lechner, L., & Wüthrich, S. (2018). Seal the deal: Bargaining positions, institutional design, and the duration of preferential trade negotiations. *International Interactions*, 44(5), 833-861. https://doi.org/10.1080/03050629.2018.1500367

Lee, A. S. (1994). Electronic mail as a medium for rich communication: An empirical investigation using hermeneutic interpretation. *MIS quarterly*, 143-157. https://doi.org/10.2307/249762

Lee, S., Lee, J., Lee, S., & Kim, J. H. (2012). Differentiating attributes in product design: A case study. *International Journal of Industrial Ergonomics*, 42(1), 15-22.

Lenzing, A., (2015). Nachverhandlungen in der Praxis: quantitative Ergebnisse einer Praktikerbefragung. *Hohenheimer Arbeits- und Projektberichte zum Marketing und Business Development*, Vol. Nr. 19.

Lewicki, R., & Litterer, J. (1985). *Negotiation*. Richard D. Irwin. Inc., Homewood, Illinois.

Lewicki, R. J., & Hiam, A. (2011). *Mastering business negotiation: a working guide to making deals and resolving conflict*. John Wiley & Sons.

Lewicki, R. J., Barry, B., & Saunders, D. M. (2015). Negotiation 7th ed. McGraw-Hill Education.

Lewis, S. A., & Fry, W. R. (1977). Effects of visual access and orientation on the discovery of integrative bargaining alternatives. *Organizational Behavior and Human Performance*, *20*(1), 75-92. https://doi.org/10.1016/0030-5073(77)90045-9

Li, H. (1996). Case-based reasoning for intelligent support of construction negotiation. *Information & Management*, *30*(5), 231-238. https://doi.org/10.1016/S0378-7206(96)01058-0

Likert, R. (1932). A technique for the measurement of attitudes. *Archives of psychology*.

Lim, J. (2003). A conceptual framework on the adoption of negotiation support systems. *Information and software technology*, *45*(8), 469-477. https://doi.org/10.1016/S0950-5849(03)00027-2

Lim, W. M., & Weissmann, M. A. (2021). Toward a theory of behavioral control. *Journal of Strategic Marketing*, 1-27. https://doi.org/10.1080/0965254X.2021.1890190

Lin, C. T., Chiu, H., & Chu, P. Y. (2006a). Agility index in the supply chain. *International Journal of production economics*, *100*(2), 285-299. https://doi.org/10.1016/j.ijpe.2004.11.013

Lin, C. T., Chiu, H., & Tseng, Y. H. (2006b). Agility evaluation using fuzzy logic. *International journal of production economics*, *101*(2), 353-368. https://doi.org/10.1016/j.ijpe.2005.01.011

Lind, M. R., & Zmud, R. W. (1991). The influence of a convergence in understanding between technology providers and users on information technology innovativeness. *Organization science*, *2*(2), 195-217. https://doi.org/10.1287/orsc.2.2.195

Lindgren, B. M., Lundman, B., & Graneheim, U. H. (2020). Abstraction and interpretation during the qualitative content analysis process. *International journal of nursing studies, 108*, 103632. https://doi.org/10.1016/j.ijnurstu.2020.103632

Linville, P. W. (1985). Self-complexity and affective extremity: Don't put all of your eggs in one cognitive basket. *Social cognition*, 3(1), 94-120. https://doi.org/10.1521/soco.1985.3.1.94

Linville, P. W. (1987). Self-complexity as a cognitive buffer against stress-related illness and depression. *Journal of personality and social psychology*, 52(4), 663. https://doi.org/10.1037/0022-3514.52.4.663

Littman, M. L., & Stone, P. (2002). *Implicit negotiation in repeated games* [Paper presentation]. Intelligent Agents VIII: Agent Theories, Architectures, and Languages 8th International Workshop*, ATAL 2001 Seattle, WA, USA, August 1–3, 2001 Revised Papers 8*, 393-404. Springer Berlin Heidelberg. https://doi.org/10.1007/3-540-45448-9_29

Liu, L. A., Chua, C. H., & Stahl, G. K. (2010). Quality of communication experience: Definition, measurement, and implications for intercultural negotiations. *Journal of Applied Psychology*, 95(3), 469. https://doi.org/10.1037/a0019094

Liu, L. A., Friedman, R., Barry, B., Gelfand, M. J., & Zhang, Z. X. (2012). The dynamics of consensus building in intracultural and intercultural negotiations. *Administrative Science Quarterly*, 57(2), 269-304. https://doi.org/10.1177/0001839212453456

Lobo, J. (2022). Virtual Physical Education: Google Meet as an alternative platform for learning skill-based concepts. *Physical education of students*, 26(6), 296-307. https://doi.10.15561/20755279.2022.0604

Loewenstein, J., & Thompson, L. L. (2006). Learning to negotiate: Novice and experienced negotiators. *Negotiation theory and research*, 77-97.

Lomas, C., Burke, M., & Page, C. L. (2008). Collaboration tools. Educause learning initiative, 2(11).

Low, K. C. P., & Anshari, M. (2013). Incorporating social customer relationship management in negotiation. *International Journal of Electronic Customer Relationship Management*, 7(3-4), 239-252. https://doi.org/10.1504/IJECRM.2013.060700

Lowery, S., & DeFleur, M. L. (1988). Milestones in mass communication research: Media effects. *(No Title).*

Lügger, K., Geiger, I., Neun, H., & Backhaus, K. (2015). When East meets West at the bargaining table: Adaptation, behavior and outcomes in intra- and intercultural German–Chinese business negotiations. *Journal of Business Economics*, 85, 15-43. https://doi.org/10.1007/s11573-013-0703-3

Lund, O. C. H., Tamnes, C. K., Moestue, C., Buss, D. M., & Vollrath, M. (2007). Tactics of hierarchy negotiation. *Journal of Research in Personality*, 41(1), 25-44. https://doi.org/10.1016/j.jrp.2006.01.002

Lutz, R. J., & Reilly, P. J. (1974). An exploration of the effects of perceived social and performance risk on consumer information acquisition. ACR North American Advances.

MacDonald, J. B., & Smith, K. (2004). The effects of technology-mediated communication on industrial buyer behavior. *Industrial Marketing Management*, 33(2), 107–116. https://doi.org/10.1016/S0019-8501(03)00033-6

Macharzina, K., & Wolf, J. (2008). *Unternehmensführung: Das internationale Managementwissen; Konzepte, Methoden, Praxis*. Springer-Verlag.

Macharzina, K., & Wolf, J. (2023). Digitalisierung und Unternehmensführung. *Unternehmensführung: Das internationale Managementwissen Konzepte–Methoden–Praxis*, 1087-1131. Wiesbaden: Springer Fachmedien Wiesbaden.

Mahenthiran, S., Greenberg, P. S., & Greenberg, R. H. (1993). The impact of computer mediated communication on the processes and outcomes of negotiated transfer pricing. *Accounting, Management and Information Technologies*, 3(4), 229–248. https://doi.org/10.1016/0959-8022(93)90019-3

Mącik, R., Mazurek, G., & Dorota, M. Ä. (2012). Channel Characteristics' Influence on Physical vs. Virtual Channel Choice for Information Search and Purchase – The Case of Polish Young Consumers. *International Journal of Cyber Society and Education*, *5*(1), 35-54.

Maity, M., & Dass, M. (2014). Consumer decision-making across modern and traditional channels: E-commerce, m-commerce, in-store. *Decision Support Systems*, *61*, 34-46.

Malhotra, D., & Bazerman, M. (2007). *Negotiation genius: How to overcome obstacles and achieve brilliant results at the bargaining table and beyond.* Bantam.

Malhotra, N. K., & Dash, S. (2011). *Marketing research an applied orientation* (paperback).

Mallen, B. (1973). Functional spin-off: A key to anticipating change in distribution structure. *Journal of Marketing*, 37(3), 18–25. https://doi.org/10.1177/002224297303700303

Mannix, E. A., Tinsley, C. H., & Bazerman, M. (1995). Negotiating over time: Impediments to integrative solutions. *Organizational Behavior and Human Decision Processes*, *62*(3), 241-251. https://doi.org/10.1006/obhd.1995.1047

Manser Payne, E., Peltier, J. W., & Barger, V. A. (2017). Omni-channel marketing, Integrated marketing communications and consumer engagement: A research agenda. *Journal of Research in Interactive Marketing*, *11*(2), 185-197. https://doi.org/10.1108/JRIM-08-2016-0091

Mansour, K., Kowalczyk, R. (2005). An Approach to One-to-Many Concurrent Negotiation. *Group Decision & Negotiation* 24, 45–66. https://doi.org/10.1007/s10726-014-9379-5

Marschak, T., & Reichelstein, S. (1998). Network mechanisms, informational efficiency, and hierarchies. *Journal of Economic Theory*, *79*(1), 106-141. https://doi.org/10.1006/jeth.1997.2375

Mars, M., Morris, C., & Scott, R. E. (2019). WhatsApp guidelines – what guidelines? A literature review. *Journal of telemedicine and telecare*, 25(9), 524-529. https://doi.org/10.1177/1357633X1987323

Marsh, H. W., & Hau, K. T. (1996). Assessing goodness of fit: Is parsimony always desirable? *Journal of experimental education*, 64(4), 364-390. https://doi.org/10.1080/00220973.1996.9944446

Martin, C. L., & Nagao, D. H. (1989). Some effects of computerized interviewing on job applicant responses. *Journal of applied psychology*, *74*(1), 72. https://doi.org/10.1037/0021-9010.74.1.72

Marquez, A. C., Bianchi, C., & Gupta, J. N. (2004). Operational and financial effectiveness of e-collaboration tools in supply chain integration. *European Journal of Operational Research*, 159(2), 348-363. https://doi.org/10.1016/j.ejor.2003.08.020

Matheson, K. (1991). Social cues in computer-mediated negotiations: Gender makes a difference. *Computers in Human Behavior*, 7(3), 137–145. https://doi.org/10.1016/0747-5632(91)90003-J

Mathews, H. L., Wilson, D. T., & Monoky Jr, J. F. (1972). Bargaining behavior in a buyer-seller dyad. *Journal of Marketing Research*, *9*(1), 103-105. https://doi.org/10.1177/002224377200900124

Matulin, M., Mrvelj, Š., Abramović, B., Šoštarić, T., & Čejvan, M. (2021, June). *User quality of experience comparison between Skype, Microsoft teams and zoom videoconferencing tools* [Paper presentation]. Future Access Enablers for Ubiquitous and Intelligent Infrastructures: 5th EAI International Conference, FABULOUS 2021, Virtual Event, May 6–7, 2021, Proceedings, 299-307. Cham: Springer International Publishing. https://doi.org/10.1007/978-3-030-78459-1_22

Maulana, M. I. (2023). Leveraging Zoom video-conferencing features in interview data generation during the Covid-19 pandemic. *Research and Teaching in a Pandemic World: The Challenges of Establishing Academic Identities During Times of Crisis*, 391-407. Singapore: Springer Nature Singapore. https://doi.org/10.1007/978-981-19-7757-2_26

Mayer, M., & Voeth, M. (2022). Improving negotiation success in B2B sales organizations: is structured negotiation management a success factor? *Journal of Business Economics*, 92, 163-196. https://doi.org/10.1007/s11573-021-01053-w

Mayfield, J., Mayfield, M., Martin, D. & Herbig, P. 1998. How location impacts international business negotiations. *Review of Business*, 19(2), 21-24.

Mayring, P. (2004). Qualitative content analysis. *A companion to qualitative research*, *1*(2), 159-176.

Mayring, P. (2015). Qualitative content analysis: Theoretical background and procedures. *Approaches to qualitative research in mathematics education: Examples of methodology and methods*, 365-380. https://doi.10.1007/978-94-017-9181-6_13

McGinn, K. L., & Croson, R. (2004). What do communication media mean for negotiations? A question of social awareness. *The handbook of negotiation and culture*, 334-349. Stanford University Press.

McGrath, J. E. (1984). *Groups: Interaction and performance* (Vol. 14). Englewood Cliffs, NJ: Prentice-Hall.

McGrath, J. E. (1993). Putting the "group" back in group support systems: Some theoretical issues about dynamic processes in groups with technological enhancements. *Group support systems: New perspectives*.

McGrath, J. E., & Hollingshead, A. B. (1994). *Groups interacting with technology: Ideas, evidence, issues, and an agenda*, ix, 181. Sage Publications, Inc.

McLuhan, M. (1964). The extension of man: Understanding media. *New York: McGraw-Hill.*

McLuhan, M. (1970). Education in the electronic age. *Interchange*, 1(4), 1–12. https://doi.org/10.1007/BF02214876

Meehl, P. E. (1970). Nuisance variables and the ex post facto design.

Meehl, P. E. (1971). High school yearbooks: A reply to Schwarz. *Journal of Abnormal Psychology, 77*(2), 143–148. https://doi.org/10.1037/h0030750

Mehrolia, S., Alagarsamy, S., Moorthy, V., & Jeevananda, S. (2023). Will Users Continue Using Banking Chatbots? The Moderating Role of Perceived Risk. *FIIB Business Review*, https://doi.10.1177.23197145231169900.

Meister, C. (2013). Kundenverhandlungen im Spiegel der internationalen Marketing Literatur. Volume 15. *Hohenheimer Arbeits- und Projektberichte zum Marketing*

Mennecke, B. E., Valacich, J. S., & Wheeler, B. C. (2000). The effects of media and task on user performance: A test of the task-media fit hypothesis. *Group Decision and Negotiation*, *9*, 507-529. https://doi.org/10.1023/A:1008770106779

Metcalf, L. E., Bird, A., Shankarmahesh, M., Aycan, Z., Larimo, J., & Valdelamar, D. D. (2006). Cultural tendencies in negotiation: A comparison of Finland, India, Mexico, Turkey, and the United States. *Journal of World Business*, 41(4), 382-394. https://doi.10.1177/1470595807079380

Milligan, G. W., & Cooper, M. C. (1985). An examination of procedures for determining the number of clusters in a data set. *Psychometrika*, 50(2), 159-179. https://doi.org/10.1007/BF02294245

Miranda, S. M., & Saunders, C. S. (2003). The social construction of meaning: An alternative perspective on information sharing. *Information systems research*, *14*(1), 87-106. https://doi.org/10.1287/isre.14.1.87.14765

Mitchell, V. W., & Greatorex, M. (1993). Risk perception and reduction in the purchase of consumer services. *Service Industries Journal*, 13(4), 179-200. https://doi.org/10.1080/02642069300000068

Mnookin, R. H., Peppet, S. R., & Tulumello, A. S. (2017). *Beyond winning: Negotiating to create value in deals and disputes.* Harvard University Press.

Moran, S., Ritov, I. (2006) Experience in Integrative Negotiations: What needs to be learned? *Journal of Experimental Social Psychology*, vol. 43(1), 77-90. https://doi.org/10.1016/j.jesp.2006.01.003

Morel, B., & Ramanujam, R. (1999). Through the looking glass of complexity: The dynamics of organizations as adaptive and evolving systems. *Organization science*, *10*(3), 278-293. https://doi.org/10.1287/orsc.10.3.278

Morris, M., Nadler, J., Kurtzberg, T., & Thompson, L. (2002). Schmooze or lose: Social friction and lubrication in e-mail negotiations. *Group Dynamics: Theory, Research, and Practice,* 6(1), 89–100. https://doi.org/10.1037/1089-2699.6.1.89

Motameni, R., & Nordstrom, R. (2014). Correlating the social media functionalities to marketing goals and strategies. *International journal of trends in marketing management*, (2). https://doi.10.15640/jmm.v2n3-4a3

Movius, H. (2008). The effectiveness of negotiation training. *Negotiation Journal*, 24(4), 509-531. https://doi.org/10.1111/j.1571-9979.2008.00201.x

Muduli, A. (2017). Workforce agility: Examining the role of organizational practices and psychological empowerment. *Global Business and Organizational Excellence*, *36*(5), 46-56. https://doi.org/10.1002/joe.21800

Nadler, J. (2004). Rapport in legal negotiation: How small talk can facilitate e-mail dealmaking. *Harvard Negotiation Literature Review*, 9, 223.

Nadler, J., Thompson, L., van Boven, L. (2003). Learning negotiation skills: Four models of knowledge creation and transfer. *Management Science*, vol. 49(4), 529-540. https://doi.org/10.1287/mnsc.49.4.529.14431

Nafei, W. A. (2016a). Organizational agility: The key to organizational success. *International Journal of Business and Management*, *11*(5), 296-309.

Nafei, W. A. (2016b). Organizational agility: The key to improve organizational performance. *International Business Research*, *9*(3), 97-111.

Naik, P. A., & Raman, K. (2003). Understanding the impact of synergy in multimedia communications. *Journal of marketing research*, *40*(4), 375-388. https://doi.org/10.1509/jmkr.40.4.375.19385

Naquin, C., & Paulson, G. (2004). Establishing trust via technology: Long distance practices and pitfalls. *International Negotiation*, 9(2), 229–244. https://doi.org/10.1163/1571806042403027

Nash Jr, J. F. (1950). The bargaining problem. *Econometrica: Journal of the econometric society*, 155-162. https://doi.org/10.2307/1907266

Neale, M. A., & Bazerman, M. H. (1991). Cognition and rationality in negotiations. *Academy of management review*, 16(1), 42-61. Free press.

Neale, M. A., & Bazerman, M. H. (1992). Negotiating rationally: the power and impact of the negotiator's frame. *Academy of Management Perspectives*, *6*(3), 42-51. https://doi.org/10.5465/ame.1992.427418

Neale, & Northcraft. (1991). Factors influencing organizational commitment. *Motivation and Work Behavior*, 5, 290–297.

Neely, A., Gregory, M., & Platts, K. (2005). Performance measurement system design: A literature review and research agenda. *International journal of operations & production management*, *25*(12), 1228-1263. https://doi.org/10.1108/01443579510083622

Neslin, S. A., & Shankar, V. (2009). Key issues in multichannel customer management: current knowledge and future directions. *Journal of interactive marketing*, *23*(1), 70-81. https://doi.10.1016/j.intmar.2008.10.005

Nijstad, & De Dreu. (2012). *Motivated information processing in organizational teams: Progress, puzzles, and prospects.* ResearchGate. http://dx.doi.org/10.1016/j.riob.2012.11.004

Nysveen, H., Pedersen, P. E., & Thorbjørnsen, H. (2005). Intentions to use mobile services: Antecedents and cross-service comparisons. *Journal of the academy of marketing science*, *33*(3), 330-346. https://doi.org/10.1177/0092070305276149

O'Brien, J. L., Martin, D. R., Heyworth, J., & Meyer, N. R. (2008). Negotiating transformational leadership: A key to effective collaboration. *Nursing & Health Sciences,* 10(2), 137-143. https://doi.org/10.1111/j.1442-2018.2008.00381.x

O'Connor, K. M., & Arnold, J. A. (2001). Distributive spirals: Negotiation impasses and the moderating role of disputant self-efficacy. *Organizational behavior and human decision processes*, *84*(1), 148-176. https://doi.org/10.1006/obhd.2000.2923

Oh, L. B., & Teo, H. H. (2006). *The impacts of information technology and managerial proactiveness in building net-enabled organizational resilience* [Paper presentation]. The Transfer and Diffusion of Information Technology for Organizational Resilience: IFIP TC8 WG 8.6 International Working Conference, June 7–10, 2006, Galway, Ireland, 33-50. Springer US. https://doi.org/10.1007/0-387-34410-1_3

Olekalns, M., Brett, J. M., & Weingart, L. R. (2003a). Phases, transitions and interruptions: Modeling processes in multi-party negotiations. *International Journal of Conflict Management*, *14*(3/4), 191-211. https://doi.org/10.1108/eb022898

Olekalns, M., & Smith, P. L. (2003b). Social motives in negotiation: The relationships between dyad composition, negotiation processes and outcomes. *International Journal of Conflict Management*, *14*(3/4), 233-254. https://doi.org/10.1108/eb022900

Park, N., Rhoads, M., Hou, J., & Lee, K. M. (2014). Understanding the acceptance of teleconferencing systems among employees: An extension of the technology acceptance model. *Computers in Human Behavior*, *39*, 118-127. https://doi.org/10.1016/j.chb.2014.05.048

Park, S., & Lee, D. (2017). An empirical study on consumer online shopping channel choice behavior in omni-channel environment. *Telematics and Informatics*, 34(8), 1398–1407. https://doi.org/10.1016/j.tele.2017.06.003

Park, S., & Park, S. (2021). How can employees adapt to change? Clarifying the adaptive performance concepts. *Human Resource Development Quarterly*, *32*(1), E1-E15. https://doi.org/10.1002/hrdq.21411

Parlamis, J., & Dibble, R. (2019). Teaming: Are two communication modes better than one?. *Team Performance Management: An International Journal*, *25*(5/6), 318-333. https://doi.org/10.1108/TPM-10-2018-0065

Parlamis, J. & Geiger, I. (2014). Is there more to email negotiation than email? The role of email affinity. Computers in Human Behavior, 32, 67–78. https://doi.org/10.1016/j.chb.2013.11.016

Parsons, S., & Wooldridge, M. (2002). Game theory and decision theory in multi-agent systems. *Autonomous Agents and Multi-Agent Systems*, *5*, 243-254. https://doi.org/10.1023/A:1015575522401

Pekkala, K., & van Zoonen, W. (2022). Work-related social media use: The mediating role of social media communication self-efficacy. *European Management Journal, 40*(1), 67-76. https://doi.org/10.1016/j.emj.2021.03.004

Pennington, A. L. (1968). Customer-salesman bargaining behavior in retail transactions. *Journal of Marketing Research*, *5*(3), 255-262. https://doi.org/10.1177/002224376800500301

Perrow, C. (1967). A framework for the comparative analysis of organizations. *American sociological review*, 194-208. https://doi.org/10.2307/2091811

Perse, E. M. (2001). *Media effects and society.* Lawrence Erlbaum Associates Publishers.

Petty, R. E., Cacioppo, J. T., Petty, R. E., & Cacioppo, J. T. (1986). *The elaboration likelihood model of persuasion* (pp. 1-24). Springer New York.

Pinfari, M. (2012). *Peace negotiations and time: deadline diplomacy in territorial disputes.* Routledge.

Pitafi, A. H., Liu, H., & Cai, Z. (2018). Investigating the relationship between workplace conflict and employee agility: The role of enterprise social media. *Telematics and Informatics*, *35*(8), 2157-2172. https://doi.org/10.1016/j.tele.2018.08.001

Plinke, W. (1985). Cost-based pricing: Behavioral aspects of price decisions for capital goods. *Journal of Business Research*, *13*(5), 447-460. https://doi.org/10.1016/0148-2963(85)90024-4

Polychroniou, A. (2014). *The SSPNet-Mobile Corpus: from the detection of non-verbal cues to the inference of social behaviour during mobile phone conversations* (Doctoral Dissertation, University of Glasgow).

Poole, M. S., & DeSanctis, G. (1990). Understanding the use of group decision support systems: The theory of adaptive structuration. *Organizations and communication technology, 173.*

Poole, M. S., Shannon, D. L., & DeSanctis, G. (1992). Communication media and negotiation processes. In L. L. Putnam & M. E. Roloff (Eds.), *Communication and negotiation*, 46–66. Sage Publications. https://doi.org/10.4135/9781483325880.n3

Pookulangara, S., & Natesan, P. (2010). Examining Consumers channel-migration intention utilizing theory of planned behavior: a multigroup analysis. *International Journal of Electronic Commerce Studies*, *1*(2), 97-116. https://dx.doi.org/10.7903/ijecs.919

Pookulangara, S., Hawley, J., & Xiao, G. (2011). Explaining multi-channel consumer's channel-migration intention using theory of reasoned action. *International Journal of Retail & Distribution Management*, *39*(3), 183-202. https://doi.org/10.1016/j.jretconser.2011.02.005

Prasarnphanich, P., & Gillenson, M. L. (2003a). The hybrid clicks and bricks business model. *Communications of the ACM*, 46(12), 178-185.

Prasarnphanich, P., & Gillenson, M. L. (2003b). The magazine archive includes every article published in Communications of the ACM for over the past 50 years. *Communications of the ACM*, 46(12), 178-185. https://doi.10.1145/953460.953498

Pruitt, D. G. (1983). Strategic choice in negotiation. *American Behavioral Scientist*, 27(2), 167-194.

Pruitt, D. G., & Carnevale, P. J. (1993). *Negotiation in social conflict*. Open University Press.

Pruitt, D. G. (1998). Social conflict. In D. T. Gilbert, S. T. Fiske, & G. Lindzey (Eds.), *The handbook of social psychology* (pp. 470–503). McGraw-Hill.

Pugh, D. S., Hickson, D. J., & Hinings, C. R. (1969). An empirical taxonomy of structures of work organizations. *Administrative science quarterly*, 115-126. https://doi.org/10.2307/2391366

Purdy, J. M., Nye, P., & Balakrishnan, P. V. (Sundar). (2000). The impact of communication media on negotiation outcomes. *International Journal of Conflict Management*, 11(2), 162–187. https://doi.org/10.1108/eb022839

Putnam, L. L., & Carcasson, M. (1997). Communication and the Oslo Negotiation: Contacts, patterns, and modes. *International Negotiation*, 2(2), 251-278. https://doi.org/10.1163/15718069720847979

Putnam, & Holmer. (1992). Framing, reframing, and issue development: Vol. SAGE *Annual Reviews of Communication Research*. SAGE Publications.

Putnam, L. L., & Jones, T. S. (1982). Reciprocity in negotiations: An analysis of bargaining interaction. *Communication monographs*, *49*(3), 171-191. https://doi.org/10.1080/03637758209376080

Putnam, L. L., & Roloff, M. E. (1992). Communication and negotiation (Vol. 20). Sage. Rice. (1984). Evaluating New Media Systems. *Evaluating the New Information Technologies.: Vol. New Directions for Program Evaluation.* Jossey-Bass.

Putnam, L.L., Phillips, N. and Chapman, P. (1996) 'Metaphors of Communication and Organization', in S.R. Clegg, C. Hardy and W. Nord (eds) *Handbook of Organizational Studies*, 375–408. London: SAGE.

Qin, R., & Nembhard, D. A. (2015). Workforce agility in operations management. *Surveys in Operations Research and Management Science*, *20*(2), 55-69. https://doi.org/10.1016/j.sorms.2015.11.001

Raab-Steiner, E., & Benesch, M. (2015). *Der Fragebogen: Von der Forschungsidee zur SPSS-Auswertung* (Vol. 8607). UTB.

Raiffa, H. (1982). The art and science of negotiation. Harvard University Press.

Raiffa, H., Richardson, J., & Metcalfe, D. D. (2002). Negotiation analysis: The science and art of collaborative decision making. Harvard University Press.

Ramayah, T. J. F. H., Cheah, J., Chuah, F., Ting, H., & Memon, M. A. (2018). Partial least squares structural equation modeling (PLS-SEM) using smartPLS 3.0. *An updated guide and practical guide to statistical analysis.* (2nd ed.). Kuala Lumpur, Malaysia: Pearson.

Reddy, M. J. (1979). The conduit metaphor: A case of frame conflict in our language about language. A. Ortony (Ed.), *Metaphor and Thought*, 164-201. Cambridge University Press.

Reisinger, T., Wagner, I., & Boiten, E. A. (2022). Security and privacy in unified communication. *ACM Computing Surveys* (CSUR), 55(3), 1-36. https://doi.org/10.1145/3498335

Rice, R. E. (1992). Task analyzability, use of new media, and effectiveness: A multi-site exploration of media richness. *Organization Science*, 3(4), 475–500. https://doi.org/10.1287/orsc.3.4.475

Rice, R. E., D'Ambra, J., & More, E. (1998). Cross-cultural comparison of organizational media evaluation and choice. *Journal of communication*, *48*(3), 3-26. https://doi.org/10.1111/j.1460-2466.1998.tb02757.x

Rigdon, E. E. (2012). Rethinking partial least squares path modeling: In praise of simple methods. *Long Range Planning, 45*(5–6), 341–358. https://doi.org/10.1016/j.lrp.2012.09.010

Rippé, C. B., Weisfeld-Spolter, S., Yurova, Y., Dubinsky, A. J., & Hale, D. (2017). Under the sway of a mobile device during an in-store shopping experience. *Psychology & Marketing*, 34(7), 733-752. https://doi.org/10.1002/mar.21019

Rogers, E. M. (1986). *Communication technology*. Simon and Schuster.

Rosette, A. S., Brett, J. M., Barsness, Z., & Lytle, A. L. (2012). When cultures clash electronically: The impact of email and social norms on negotiation behavior and outcomes. *Journal of Cross-Cultural Psychology*, 43(4), 628–643. https://doi.org/10.1177/0022022111407190

Ross, W., & LaCroix, J. (1996). Multiple meanings of trust in negotiation theory and research: A literature review and integrative model. *International Journal of Conflict Management*, 7(4), 314-360. https://doi.org/10.1108/eb022786

Rossiter, J. R. (2008). Content validity of measures of abstract constructs in management and organizational research. *British Journal of Management*, 19(4), 380-388. https://doi.org/10.1111/j.1467-8551.2008.00587.x

Ryoo, S. Y., & Koo, C. (2010). The moderating effect of media synchronicity in the communication media use and knowledge creation. *Asia Pacific Journal of Information Systems*, *20*(2), 103-124.

Salacuse, J. W. (2001). Renegotiating existing agreements: How to deal with "life struggling against form". *Negotiation Journal*, 17(4), 311-331. https://doi.org/10.1111/j.1571-9979.2001.tb00243.x

Salacuse, J. W. (2005). *The global negotiator: Making, managing, and mending deals around the world in the twenty-first century.* Palgrave Macmillan.

Salmon, E. D., Gelfand, M. J., Çelik, A. B., Kraus, S., Wilkenfeld, J., & Inman, M. (2013). Cultural contingencies of mediation: Effectiveness of mediator styles in intercultural disputes. *Journal of Organizational Behavior*, 34(6), 887-909. https://doi.org/10.1002/job.1870

Sarstedt, M., Ringle, C. M., & Hair, J. F. (2021). Partial least squares structural equation modeling. *Handbook of market research* (pp. 587-632). Cham: Springer International Publishing. https://doi.org/10.1007/978-3-319-57413-4_15

Sawyer, J., & Guetzkow, H. S. (1965). *Bargaining and negotiation in international relations*. H. Rinehart and Winston.

Schmidt, M. (2020). *Fragmentierung von Verhandlungen* (Doctoral Dissertation). Cumulative thesis in Negotiation Management. Publisher Dr. Kovac. University of Hohenheim, Stuttgart.

Schmidtke, J. M., & Cummings, A. (2017). The effects of virtualness on teamwork behavioral components: The role of shared mental models. *Human Resource Management Review*, *27*(4), 660-677. https://doi.org/10.1016/j.hrmr.2016.12.011

Scholz, J. P., & Schöner, G. (1999). The uncontrolled manifold concept: identifying control variables for a functional task. *Experimental brain research*, *126*, 289-306.

Schoop, M., Köhne, F., Staskiewicz, D., Voeth, M., & Herbst, U. (2008). The antecedents of renegotiations in practice — an exploratory analysis. *Group Decision and Negotiation*, 17, 127-139. https://doi.org/10.1007/s10726-007-9080-z

Schoop, M. (2010). Support of complex electronic negotiations. *Handbook of group decision and negotiation*, 409-423. https://doi.10.1007/978-90-481-9097-3_24

Schoop, M. (2021a). Negoisst: complex digital negotiation support. *Handbook of group decision and negotiation*, 1149-1167. https://doi.10.1007/978-3-030-49629-6_24

Schoop, M. (2021b). Negotiation communication revisited. *Central European Journal of Operations Research*, 29(1), 163-176. https://doi.org/10.1007/s10100-020-00730-5

Schwarz, S. (2012). *Analyse und Steuerung von Verhandlungsperformance in industriellen Verhandlungen* (Doctoral Dissertation). Cumulative thesis in Negotiation Management. Publisher Dr. Kovač. University of Hohenheim, Stuttgart.

Sheppard, J. M., & Young, W. B. (2006). Agility literature review: Classifications, training and testing. *Journal of sports sciences*, *24*(9), 919-932. https://doi.org/10.1080/02640410500457109

Selvi, A. F. (2019). Qualitative content analysis. In *The Routledge handbook of research methods in applied linguistics*, 440-452. Routledge.

Senge, P. M. (1994). *The fifth discipline fieldbook: Strategies and tools for building a learning organization*. Currency.

Seyff, N., Todoran, I., Caluser, K., Singer, L., & Glinz, M. (2015). Using popular social network sites to support requirements elicitation, prioritization and negotiation. *Journal of Internet Services and Applications*, 6(1), 1-16. https://doi.org/10.1186/s13174-015-0021-9

Shafer, R. A. (1997). *Creating organizational agility: The human resource dimension*. Cornell University.

Shahriza, A. K. N., & Heckman, R. (2005). Group communication media choice and the use of information and communication technology to support learning: A case study. *Campus-Wide Information Systems*, 22(1), 28–42. https://doi.org/10.1108/10650740510574401

Shakun, M. F. (2005). Multi-bilateral multi-issue e-negotiation in e-commerce with a tit-for-tat computer agent. *Group Decision and Negotiation*, 14(5), 383–392. https://doi.org/10.1007/s10726-005-8353-7

Shannon, C. E. (1949). Communication in the presence of noise. *Proceedings of the IRE*, *37*(1), 10-21.

Shao, B., Wang, L., Cheng, D., & Doucet, L. (2015). Anger suppression in negotiations: The roles of attentional focus and anger source. Journal of Business and Psychology, 30, 747-758.

Sharifi, H., & Zhang, Z. (1999). A methodology for achieving agility in manufacturing organisations: An introduction. *International journal of production economics*, *62*(1-2), 7-22.
https://doi.org/10.1016/S0925-5273(98)00217-5

Sharifi, H., & Zhang, Z. (2001). Agile manufacturing in practice-Application of a methodology. *International journal of operations & production management*, *21*(5/6), 772-794. https://doi.org/10.1108/01443570110390462

Sharifi, H., Colquhoun, G., Barclay, I., & Dann, Z. (2001). Agile manufacturing: a management and operational framework. *Proceedings of the Institution of Mechanical engineers, Part B: Journal of engineering Manufacture*, *215*(6), 857-869. https://doi.org/10.1243/0954405011518647

Sheer, V. C., & Rice, R. E. (2017). Mobile instant messaging use and social capital: Direct and indirect associations with employee outcomes. *Information & Management*, 54(1), 90-102. https://doi.org/10.1016/j.im.2016.04.001

Sheffield, J. (1995). The effect of communication medium on negotiation performance. *Group Decision and Negotiation*, *4*, 159-179.
https://doi.org/10.1007/BF01410100

Shell, G. R. (2006). *Bargaining for advantage: Negotiation strategies for reasonable people.* Penguin.

Sherehiy, B., Karwowski, W., & Layer, J. K. (2007). A review of enterprise agility: Concepts, frameworks, and attributes. *International Journal of industrial ergonomics*, *37*(5), 445-460. https://doi.org/10.1016/j.ergon.2007.01.007

Sherehiy, B. (2008). *Relationships between agility strategy, work organization and workforce agility.* University of Louisville.

Sheu, J. B. (2016). Buyer behavior in quality-dominated multi-sourcing recyclable-material procurement of green supply chains. *Production and Operations Management*, 25(3), 477-497. https://doi.org/10.1111/poms.12263

Shin, I. (2014). *Factors Influencing the Cooperative Process of Cross-border Merger and Acquisitions Negotiation in Face-to-Face: The Role of Individual, Strategic, and Contextual Components* (Doctoral Dissertation, King's College London).

Short, J. A. (1974). Effects of medium of communication on experimental negotiation. *Human Relations*, 27(3), 225–234. https://doi.org/10.1177/001872677402700303

Short, Williams, & Christie. (1976). The social psychology of telecommunications. Wiley.

Shmueli, G., & Koppius, O. R. (2011). Predictive analytics in information systems research. *MIS Quarterly, 35*(3), 553–572. https://doi.org/10.2307/23042796

Shmueli, G., Ray, S., Estrada, J. M. V., & Chatla, S. B. (2016). The elephant in the room: predictive performance of PLS models. *Journal of Business Research*, 69(10), 4552-4564. https://doi.org/10.1016/j.jbusres.2016.03.049

Siegel, J., Dubrovsky, V., Kiesler, S., & McGuire, T. W. (1986). Group processes in computer-mediated communication. *Organizational behavior and human decision processes*, *37*(2), 157-187. https://doi.org/10.1016/0749-5978(86)90050-6

Sillars, A. L. (1980). The sequential and distributional structure of conflict interactions as a function of attributions concerning the locus of responsibility and stability of conflicts. *Annals of the International Communication Association*, *4*(1), 217-235. https://doi.org/10.1080/23808985.1980.11923804

Sillars, A. L., Coletti, S. F., Parry, D., & Rogers, M. A. (1982). Coding verbal conflict tactics: Nonverbal and perceptual correlates of the "avoidance-distributive-integrative" distinction. *Human communication research*, *9*(1), 83-95. https://doi.org/10.1111/j.1468-2958.1982.tb00685.x

Simonelli, N. M. (2011). Bargaining over international multilateral agreements: The duration of negotiations. *International Interactions*, 37(2), 147-169. https://doi.org/10.1080/03050629.2011.568845

Sitkin, S. B., Sutcliffe, K. M., & Barrios-Choplin, J. R. (1992). A dual-capacity model of communication media choice in organizations. *Human Communication Research*, 18(4), 563–598. https://doi.org/10.1111/j.1468-2958.1992.tb00572.x

Spears, R., & Lea, M. (1994). Panacea or panopticon? The hidden power in computer-mediated communication. *Communication Research*, 21(4), 427–459. https://doi.org/10.1177/009365094021004001

Spector, P. E., Chen, P. Y., & O'Connell, B. J. (2000). A longitudinal study of relations between job stressors and job strains while controlling for prior negative affectivity and strains. Journal of Applied Psychology, 85(2), 211. https://doi.org/10.1037/0021-9010.85.2.211

Spector, P. E., & Brannick, M. T. (2011). Methodological urban legends: The misuse of statistical control variables. Organizational research methods, 14(2), 287-305, https://doi.10.1177/1094428110369842.

Sproull, L., & Kiesler, S. (1986). Reducing social context cues: Electronic mail in organizational communications. Management Science, 32(11), 1492–1512. JSTOR.

Staples, D. S., & Zhao, L. (2006). The effects of cultural diversity in virtual teams versus face-to-face teams. Group Decision and Negotiation, 15(4), 389–406. https://doi.org/10.1007/s10726-006-9042-x

Stevens, C. K., & Gist, M. E. (1997). Effects of self-efficacy and goal-orientation training on negotiation skill maintenance: What are the mechanisms? Personnel psychology, 50(4), 955-978.

Stone, R. N., & Grønhaug, K. (1993). Perceived risk: Further considerations for the marketing discipline. European Journal of marketing, 27(3), 39-50. https://doi.org/10.1108/03090569310026637

Stoshikj, M. (2014). Integrative and distributive negotiations and negotiation behavior. Journal of Service Science Research, 6, 29-69. https://doi.org/10.1007/s12927-014-0002-8

Straus, S. G., & McGrath, J. E. (1994). Does the medium matter? The interaction of task type and technology on group performance and member reactions. *Journal of applied psychology*, 79(1), 87. https://doi.org/10.1037/0021-9010.79.1.87

Stuhlmacher, A. F., & Citera, M. (2005). Hostile behavior and profit in virtual negotiation: A meta-analysis. *Journal of Business and Psychology*, 20(1), 69–93. https://doi.org/10.1007/s10869-005-6984-y

Stuhlmacher, A. F., Citera, M., & Willis, T. (2007). Gender differences in virtual negotiation: Theory and research. *Sex Roles*, 57(5–6), 329–339. https://doi.org/10.1007/s11199-007-9252-y

Suh, K. S. (1999). Impact of communication medium on task performance and satisfaction: An examination of media-richness theory. *Information & Management*, 35(5), 295–312. https://doi.org/10.1016/S0378-7206(98)00097-4

Sullivan, B. A., O'Connor, K. M., & Burris, E. R. (2006). Negotiator confidence: The impact of self-efficacy on tactics and outcomes. *Journal of Experimental Social Psychology*, *42*(5), 567-581. https://doi.org/10.1016/j.jesp.2005.09.006

Sumukadas, N., & Sawhney, R. (2004). Workforce agility through employee involvement. *Iie Transactions*, *36*(10), 1011-1021. https://doi.org/10.1080/07408170490500997

Swaab, R. I., Galinsky, A. D., Medvec, V., & Diermeier, D. A. (2012). The Communication Orientation Model: Explaining the diverse effects of sight, sound, and synchronicity on negotiation and group decision-making outcomes. *Personality and Social Psychology Review*, 16(1), 25–53. https://doi.org/10.1177/1088868311417186

Tarafdar, M., Tu, Q., & Ragu-Nathan, T. S. (2010). Impact of technostress on end-user satisfaction and performance. *Journal of management information systems*, 27(3), 303-334. https://doi.org/10.2753/MIS0742-1222270311

Taylor, S., & Todd, P. (1995). Decomposition and crossover effects in the theory of planned behavior: A study of consumer adoption intentions. *International journal of research in marketing*, *12*(2), 137-155. https://doi.org/10.1016/0167-8116(94)00019-K

Te'Eni, D. (2001). A cognitive-affective model of organizational communication for Designing IT. *MIS quarterly*, 251-312. https://doi.org/10.2307/3250931

Terry, D. J., & O'Leary, J. E. (1995). The theory of planned behaviour: The effects of perceived behavioural control and self-efficacy. *British journal of social psychology*, *34*(2), 199-220. https://doi.org/10.1111/j.2044-8309.1995.tb01058.x

Thompson, J. D. (1967). *Organizations in action: Social science bases of administrative theory.* New York: McGraw-Hill.

Thompson, L., & Hastie, R. (1990). Social perception in negotiation. *Organizational Behavior and Human Decision Processes*, 47(1), 98-123. https://doi.org/10.1016/0749-5978(90)90048-E

Thompson, L. (1990). Negotiation behavior and outcomes: Empirical evidence and theoretical issues. *Psychological bulletin*, 108(3), 515. https://doi.org/10.1037/0033-2909.108.3.515

Thompson, L., & DeHarpport, T. (1998). Relationships, goal incompatibility, and communal orientation in negotiations. *Basic and applied social psychology*, 20(1), 33-44. https://doi.org/10.1016/j.obhdp.2016.06.003

Thompson, R. L. (2005). *The US Farm Bill and the Doha Negotiations: On Parallel Tracks or a Collision Course?* [Paper presentation] IPC Issue Brief, International Food and Agricultural Trade Policy Council. https://www. agritrade. org/Publications/iBs/329701.pdf

Thompson, L. L., Wang, J., & Gunia, B. C. (2010). Negotiation. *Annual review of psychology*, 61, 491-515.

Thompson, L. (2011). *The mind and heart of the negotiator.* Prentice Hall.

Tinsley, C. H., & Brett, J. M. (2001). Managing workplace conflict in the United States and Hong Kong. *Organizational behavior and human decision processes*, 85(2), 360-381. https://doi.org/10.1006/obhd.2000.2944

Trenz, M. (2015). *Multichannel commerce: A consumer perspective on the integration of physical and electronic channels.* Springer.

Trevino, L. K., Lengel, R. H., & Daft, R. L. (1987). Media symbolism, media richness, and media choice in organizations: A symbolic interactionist perspective. *Communication Research*, 14(5), 553–574. https://doi.org/10.1177/009365087014005006

Trevino, L. K., Daft, R. L., & Lengel, R. H. (1990). Understanding Managers' Media Choices: A Symbolic Interactionist. *Organizations and communication technology, 71*.

Trizano-Hermosilla, I., & Alvarado, J. M. (2016). Best alternatives to Cronbach's alpha reliability in realistic conditions: congeneric and asymmetrical measurements. *Frontiers in psychology*, *7*, 769. https://doi.org/10.3389/fpsyg.2016.00769

Tsfati, Y., Cohen, J., & Gunther, A. C. (2011). The influence of presumed media influence on news about science and scientists. *Science Communication*, *33*(2), 143-166. https://doi.org/10.1177/1075547010380385

Tushman, M. L., & Nadler, D. A. (1978). Information processing as an integrating concept in organizational design. *Academy of management review*, 3(3), 613-624. https://doi.org/10.5465/amr.1978.4305791

Valacich, J. S., Dennis, A. R., & Connolly, T. (1994). Idea generation in computer-based groups: A new ending to an old story. *Organizational behavior and human decision processes*, 57(3), 448-467. https://doi.org/10.1006/obhd.1994.1024

Vaismoradi, M., & Snelgrove, S. (2019). Theme in qualitative content analysis and thematic analysis. *Qualitative Social Research* (Vol. 20, No. 3). https://nbn-resolving.org/urn:nbn:de:0168-ssoar-65383-7

Van de Ven, A. H. (1976). On the nature, formation, and maintenance of relations among organizations. *Academy of management review, 1*(4), 24-36. https://doi.org/10.5465/amr.1976.4396447

Van de Ven, A. H. (1985). Spinning on Symbolism: The Problem of Ambivalence. *Journal of Management* 11: 101-2.

Veiga de Paula, M. M., de Souza, J. M., Blaschek, J. R., & Baião, F. (2004, January). A Cooperative System to Support Inventory Leveling Negotiations. In *Proceedings of the 37th Annual Hawaii International Conference on System Sciences (HICSS'04)-Track 7-Volume 7* (pp. 70192-2).

Verhoef, P. C., Neslin, S. A., & Vroomen, B. (2007). Multichannel customer management: Understanding the research-shopper phenomenon. *International Journal of Research in Marketing*, 24(2), 129-148. https://doi.org/10.1016/j.ijresmar.2006.11.002

Verhoef, P. C., Kannan, P. K., & Inman, J. J. (2015). From multi-channel retailing to omni-channel retailing. *Journal of Retailing*, 91(2), 174–181. https://doi.org/10.1016/j.jretai.2015.02.005

Vinodh, S., & Prasanna, M. (2011). Evaluation of agility in supply chains using multi-grade fuzzy approach. *International Journal of Production Research*, *49*(17), 5263-5276. https://doi.org/10.1080/00207543.2010.524260

Vinzi, V. E., Chin, W. W., Henseler, J., & Wang, H. (2010). *Handbook of partial least squares* (Vol. 201, No. 0). Berlin: Springer.

Voeth, M., Gawantka, A., & Herbst, U. (2006, May). The analysis of negotiation satisfaction. In Proceedings of the European Marketing Academy Conference (EMAC 2006), Athens, Greece.

Voeth, M., & Becker, T. (2013). Der Einkäufer als Verhandlungspartner – Eine empirische Studie Nr. 13 *Hohenheim Arbeits- und Projektberichte zum Marketing.*

Voeth, M., & Lenzing, A. C. (2017). Renegotiations – empirical analysis of impacts on business relationships. *Journal of Business & Industrial Marketing*, 32(4), 541-552. https://doi.org/10.1108/JBIM-09-2015-0175

Voeth, M., Herbst, U., & Stief, S. (2015). *Wie verhandelt die Praxis? Ergebnisse einer Befragung von deutschen Managern.* Publisher Negotiation Academy Potsdam, Working Paper No. 1, Stuttgart 2015.

Voeth, M., & Herbst, U. (2015). *Verhandlungsmanagement: Planung, Steuerung und Analyse.* Schäffer-Poeschel.

Voeth, M., Herbst, U., Haggenmüller, S., & Weber, M. (2018). *Wie verhandeln deutsche Manager?* Publisher Negotiation Academy Potsdam, Working Paper No. 3, Stuttgart 2018.

Voeth, M., & Mayer, M. (2018). Ansätze für die Institutionalisierung von Verhandlungsmanagement im Vertrieb. *Die Unternehmung* 72(1), 87–106. https://www.jstor.org/stable/26610466

Voeth, M., Herbst, U., Haggenmüller, S., & Weber, M. C. (2020). Wie verhandeln deutsche Manager? *Zeitschrift für Konfliktmanagement*, 23(1), 21-24. https://doi.org/10.9785/zkm-2020-230107

Voeth, M., & Herbst, U. (2022). Price Negotiations in Commodity Markets. In Commodity Marketing: Strategies, *Concepts, and Cases*, 85-103. Cham: Springer International Publishing.
https://doi.org/10.1007/978-3-030-90657-3_5

Voeth, M., Herbst, U., Urbitsch, Y., & Oryl, M. (2022). Wie verhandeln deutsche Manager nach der Pandemie? *Negotiation Academy Potsdam*, Working Paper No. 4, Stuttgart.

Volkema, R., Kapoutsis, I., & Nikolopoulos, A. (2013). Initiation behavior in negotiations: The moderating role of motivation on the ability–intentionality relationship. *Negotiation and Conflict Management Research*, 6(1), 32-48. https://doi.org/10.1111/ncmr.12002

Vroom, V. H. (1964). *Work and motivation*. Wiley.

Walther. (1992). Interpersonal effects in computer-mediated interaction: A relational perspective. *Communication Research*, 19(1), 52–90. https://doi.org/10.1177/009365092019001003

Walther. (1994). Anticipated ongoing interaction versus channel effects on relational communication in computer-mediated interaction. *Human Communication Research*, 20(4), 473–501. https://doi.org/10.1111/j.1468-2958.1994.tb00332.x

Walther. (2015). *Engaging theories in interpersonal communication: Social Information Processing Theory.* SAGE Publications.

Wang, A. (2006). Advertising engagement: A driver of message involvement on message effects. Journal of advertising research, 46(4), 355-368. https://doi.10.2501/S0021849906060429

Wang, H.-C., & Doong, H.-S. (2014). Revisiting the task-media fit circumflex: A further examination of negotiation tasks. *Information & Management*, 51(6), 738–746. https://doi.org/10.1016/j.im.2014.07.001

Wang, C. X., Ghazal, A., Ai, B., Liu, Y., & Fan, P. (2015). Channel measurements and models for high-speed train communication systems: A survey. *IEEE communications surveys & tutorials*, *18*(2), 974-987. https://doi.10.1109/COMST.2015.2508442

Wang, Y. M., Lin, H. H., Tai, W. C., & Fan, Y. L. (2016). Understanding multichannel research shoppers: an analysis of Internet and physical channels. *Information Systems and e-Business Management*, *14*, 389-413.

Wangenheim, F. V., & Bayón, T. (2007). Behavioral consequences of overbooking service capacity. *Journal of Marketing*, *71*(4), 36-47. https://doi.org/10.1509/jmkg.71.4.036

Warner, R. (2008). *Applied statistics from bivariate through multivariate techniques*. Los Angeles: Sage publications.

Watkins, M. (1999). Negotiating in a complex world. *Negotiation Journal*, 15(3), 245-270. https://doi.org/10.1111/j.1571-9979.1999.tb00195.x

Watkins, M. D. (2014). *Negotiation excellence: Successful deal making*. John Wiley & Sons.

Watson-Manheim, M. B., & Bélanger, F. (2007). Communication media repertoires: Dealing with the Multiplicity of Media Choices. *MIS Quarterly*, 31(2), 267–293. JSTOR. https://doi.org/10.2307/25148791

Weber, M. C. (2019). *Performance Enhancing Strategies in Different Negotiation Phases* (Doctoral Dissertation. Cumulative thesis in Negotiation Management. Publisher Verlag Dr. Kovac. University of Potsdam.

Weick, K. E. (2015). Karl E. WEICK (1979), The Social Psychology of Organizing, Paperback: 294 pages Publisher: McGraw-Hill (1979) Language: English ISBN: 978-0075548089. *M@ n@ gement*, (2), 189-193. https://doi.org/10.3917/mana.182.0189

Weingart, L. R., Bennett, R. J., & Brett, J. M. (1993). The impact of consideration of issues and motivational orientation on group negotiation process and outcome. *Journal of Applied Psychology*, *78*(3), 504. https://doi.org/10.1037/0021-9010.78.3.504

Weisberg, H. F. (2008). The methodological strengths and weaknesses of survey research. Sage Publications.

Weisberg, D. S., Hirsh-Pasek, K., & Golinkoff, R. M. (2013). Embracing complexity: Rethinking the relation between play and learning: Comment on Lillard et al. (2013). *Psychological Bulletin, 139*(1), 35–39. https://doi.org/10.1037/a0030077

Weiss, S. E. (1997). Explaining outcomes of negotiation: toward a grounded model for negotiation between organizations. Research on negotiation in organizations, 6, 247-333.

Weiss, S. (2006). International business negotiation in a globalizing world: Reflections on the contributions and future of a (sub) field. *International Negotiation*, 11(2), 287–316. https://doi.org/10.1163/157180606778968371

Wellens, A. R. (1986). Use of a psychological distancing model to assess differences in telecommunication media. *Teleconferencing and electronic media, 5*, 347-361.

Wellens, A. R. (1989, May). Effects of telecommunication media upon information sharing and team performance: Some theoretical and empirical observations. In *Proceedings of the IEEE National Aerospace and Electronics Conference* (pp. 726-733). IEEE. https://doi.10.1109/NAECON.1989.40291

Westmyer, S. A., DiCioccio, R. L., & Rubin, R. B. (1998). Appropriateness and effectiveness of communication channels in competent interpersonal communication. *Journal of communication*, *48*(3), 27-48. https://doi.org/10.1111/j.1460-2466.1998.tb02758.x

Wilcox, K., Kramer, T., & Sen, S. (2011). Indulgence or self-control: A dual process model of the effect of incidental pride on indulgent choice. *Journal of Consumer Research*, *38*(1), 151-163.

Williams, E. (1977). Experimental comparisons of face-to-face and mediated communication: A review. *Psychological bulletin*, *84*(5), 963. https://doi.org/10.1037/0033-2909.84.5.963

Wood, R. E. (1986). Task complexity: Definition of the construct. *Organizational behavior and human decision processes*, *37*(1), 60-82. https://doi.org/10.1016/0749-5978(86)90044-0

Woodside, A. G., & Davenport Jr, J. W. (1974). The effect of salesman similarity and expertise on consumer purchasing behavior. *Journal of Marketing Research*, *11*(2), 198-202. https://doi.org/10.1177/002224377401100212

Woodward, J. (1965). *Industrial Organization: Theory and Practice*. Oxford University Press

Yang, G. (2012). Analyzing computer mediated buyer-seller negotiations: an application of social presence theory. *International Journal of Marketing & Business Communication*, *1*(4), 25.

Yates, J., & Orlikowski, W. J. (1992). Genres of organizational communication: A structurational approach to studying communication and media. *Academy of management review*, 17(2), 299-326. https://doi.org/10.5465/amr.1992.4279545

Yip, J. A., & Schweinsberg, M. (2017). Infuriating impasses: Angry expressions increase exiting behavior in negotiations. *Social Psychological and Personality Science*, 8(6), 706-714. https://doi.org/10.1177/1948550616683021

Yu, U. J., Niehm, L. S., & Russell, D. W. (2011). Exploring perceived channel price, quality, and value as antecedents of channel choice and usage in multi-channel shopping. *Journal of Marketing Channels*, *18*(2), 79-102. https://doi.org/10.1080/1046669X.2011.558826

Yuan, Y., Head, M., & Du, M. (2003). The effects of multimedia communication on web-based negotiation. *Group Decision and Negotiation*, *12*, 89-109. https://doi.org/10.1023/A:1023016804379

Yzer, M. (2012). Perceived behavioral control in reasoned action theory: A dual-aspect interpretation. *The annals of the American academy of political and social science*, *640*(1), 101-117. https://doi.org/10.1177/0002716211423500

Zallocco, R., Bolman Pullins, E., & Mallin, M. L. (2009). A re-examination of B2B sales performance. *Journal of Business & Industrial Marketing*, 24(8), 598-610. https://doi.org/10.1108/08858620910999466

Zartman, I. W. (1988). Negotiations in South Africa. *The Washington Quarterly*, 11(4), 141-158. https://doi.org/10.1080/01636608809477507

Zartman, I. W. (1994). *International multilateral negotiation: Approaches to the management of complexity*. Jossey-Bass.

Zartman, I. W. (2003). Conclusion: Managing Complexity. *International Negotiation*, 8(1), 179–186. https://doi.org/10.1163/138234003769590712

Zartman, I. W., & Berman, M. R. (1982). *The practical negotiator*. Yale University Press.

Zhang, S., Makedon, F., Ford, J., Sudborough, C., Ai, L., Kapidakis, S., ... & Loukis, E. (2004, August). An international trade negotiation framework for e-government. In *International Conference on Electronic Government* (pp. 211-217). Berlin, Heidelberg: Springer Berlin Heidelberg. https://doi.org/10.1007/978-3-540-30078-6_35

Ziegele, F. (2006). Erfolgreich verhandeln. Bedingungen effizienter Zielvereinbarungen. *Die Hochschule: Journal für Wissenschaft und Bildung, 15*(2), 18-33.

Zornoza, A., Ripoll, P., & Peiró, J. M. (2002). Conflict management in groups that work in two different communication contexts: Face-to-face and computer-mediated communication. *Small Group Research*, 33(5), 481–508. https://doi.org/10.1177/104649602237167

Bibliography

Acker, S. R., & Levitt, S. R. (1987). Designing videoconference facilities for improved eye contact. *Journal of Broadcasting & Electronic Media*, 31(2), 181–191. https://doi.org/10.1080/08838158709386656

Adair, W. L., & Brett, J. M. (2005). The negotiation dance: Time, culture, and behavioral sequences in negotiation. *Organization Science*, 16(1), 33-51. https://doi.org/10.1287/orsc.1040.0102

Aguinis, H., Edwards, J. R., & Bradley, K. J. (2016). Improving our understanding of moderation and mediation in strategic management research. *Organizational Research Methods*. https://doi.1094428115627498

Aguinis, H., Edwards, J. R., & Bradley, K. J. (2017). Improving our understanding of moderation and mediation in strategic management research. *Organizational Research Methods*, *20*(4), 665-685. https://doi.org/10.1177/1094428115627498

Ajzen, I., & Madden, T. J. (1986). Prediction of goal-directed behavior: Attitudes, intentions, and perceived behavioral control. *Journal of experimental social psychology*, *22*(5), 453-474. https://doi.org/10.1016/0022-1031(86)90045-4

Argyle, M. (2013). *Bodily Communication*. Routledge.

Arunachalam, V., & Dilla, W. N. (1995). Judgment accuracy and outcomes in negotiation: A causal modeling analysis of decision-aiding effects. *Organizational Behavior and Human Decision Processes*, 61(3), 289–304. https://doi.org/10.1006/obhd.1995.1023

Bichler, M., Kersten, G., & Strecker, S. (2003). Towards a structured design of electronic negotiations. *Group Decision and Negotiation*, *12*, 311-335. https://doi.org/10.1023/A:1024867820235

Bodensteiner, W. D. (1970). *Information channel utilization under varying research and development project conditions: An aspect of inter-organizational communication channel usage*. The University of Texas at Austin.

Cai, X., Zhang, Y., & Zhou, L. (2019). Research on multi-channel negotiation decision-making based on clustering algorithm. *Journal of Ambient Intelligence and Humanized Computing*, 10(11), 4373-4383.

Cameron, D. (2008). *The myth of Mars and Venus: Do men and women really speak different languages?* Oxford University Press.

Cavana, R. Y., Delahaye, B. L., & Sekaran, U. (2001). *Applied business research: Qualitative and quantitative methods.* John Wiley & Sons.

Cheah, J. H., Ting, H., Thaichon, P., Fam. K.-S., & Bazylewich, M. (2020). Can positioning strategies help influence willingness to pay for office space? Evidence on the moderating effect of office space grade and industry sector for occupiers of leased office space, *Journal of Strategic Marketing.* https://doi.10.1080/0965254X.2020.1733049

Chen, Y., Chen, X., & Chen, Y. (2021). Multi-channel negotiation strategy based on cluster analysis. *Journal of Intelligent & Fuzzy Systems*, 41(2), 2723-2734.

Chen, S., & Weiss, G. (2013). An efficient automated negotiation strategy for complex environments. *Engineering Applications of Artificial Intelligence*, 26(10), 2613–2623. https://doi.org/10.1016/j.engappai.2013.08.012

Chin, W. W., Marcolin, B. L., & Newsted, P. R. (1996). *A partial least squares latent variable modeling approach for measuring interaction effects: Results from a Monte Carlo simulation study and voice mail emotion/adoption study* [Paper presentation]. In J. I. DeGross, S. L. Jarvenpaa, & A. Srinivasan (Eds.), Proceedings of the Seventeenth International Conference on Information Systems, 21–41. Cleveland, OH: Association for Information Systems.

Chin, W. W. (1998). The partial least squares approach to structural equation modeling. *Modern methods for business research*, 295(2), 295-336.

Cohen, J. (1978). Partialed products are interactions; partialed powers are curve components. *Psychological Bulletin*, *85*(4), 858. https://doi.org/10.1037/0033-2909.85.4.858

Curhan, J. R., Elfenbein, H. A., & Xu, H. (2006). What do people value when they negotiate? Mapping the domain of subjective value in negotiation. *Journal of Personality and Social Psychology*, 91(3), 493-512. https://doi.org/10.1037/0022-3514.91.3.493

Curhan, J. R., Elfenbein, H. A., & Eisenkraft, N. (2010). The objective value of subjective value: A multi-round negotiation study. *Journal of Applied Social Psychology*, *40*(3), 690-709. https://doi.org/10.1111/j.1559-1816.2010.00593.x

Daft, R. L., & Macintosh, N. B. (1981). A tentative exploration into the amount and equivocality of information processing in organizational work units. *Administrative Science Quarterly*, 26(2), 207–224. JSTOR. https://doi.org/10.2307/2392469

Echambadi, R., & Hess, J. D. (2007). Mean-centering does not alleviate collinearity problems in moderated multiple regression models. *Marketing Science*, *26*(3), 438-445. https://doi.org/10.1287/mksc.1060.0263

Falk, R. F., & Miller, N. B. (1992). *A primer for soft modeling*. University of Akron Press.

Fassott, G., Henseler, J., & Coelho, P. S. (2016). Testing moderating effects in PLS path models with composite variables. *Industrial Management & Data Systems*, *116*(9). https://doi.org/10.1108/IMDS-06-2016-0248

Finney, J. W., Mitchell, R. E., Cronkite, R. C., & Moos, R. H. (1984). Methodological issues in estimating main and interactive effects: Examples from coping/social support and stress field. *Journal of Health and Social behavior*, 85-98. https://doi.org/10.2307/2136706

Floyd, F. J., & Widaman, K. F. (1995). Factor analysis in the development and refinement of clinical assessment instruments. *Psychological assessment*, 7(3), 286–299. https://doi.org/10.1037/1040-3590.7.3.286

Fraley, C., & Raftery, A. E. (1998). How many clusters? Which clustering method? Answers via model-based cluster analysis. *The Computer Journal*, 41(8), 578-588. https://doi.org/10.1093/comjnl/41.8.578

Galin, A., Gross, M., & Gosalker, G. (2007). E-negotiation versus face-to-face negotiation what has changed – if anything? *Computers in Human Behavior*, 23(1), 787–797. https://doi.org/10.1016/j.chb.2004.11.009

Giordano, G. A., Stoner, J. S., Brouer, R. L., & George, J. F. (2007). The influences of deception and computer-mediation on dyadic negotiations. *Journal of Computer-Mediated Communication*, 12(2), 362–383. https://doi.org/10.1111/j.1083-6101.2007.00329.x

Gelfand, M. J., & Brett, J. M. (2004). *The Handbook of Negotiation and Culture*. Stanford University Press.

Gibbs, R. W. (1994). *The poetics of mind: Figurative thought, language, and understanding*. Cambridge University Press.

Graham, J. L., & Mintu-Wimsat, A. (1997). Culture's influence on business negotiations in four countries. *Group Decision and Negotiation*, *6*, 483-502. https://doi.org/10.1023/A:1008655003951

Gujarati, D. (1970). Use of dummy variables in testing for equality between sets of coefficients in two linear regressions: a note. *The American Statistician*, *24*(1), 50-52. https://doi.10.1080/00031305.1970.10477181

Graf, A., Koeszegi, S. T., & Pesendorfer, E. (2010). Electronic negotiations in intercultural interfirm relationships. *Journal of Managerial Psychology*, 25(5), 495–512. https://doi.org/10.1108/02683941011048391

Hair, J. F., Ringle, C. M., & Sarstedt, M. (2011). PLS-SEM: Indeed, a silver bullet. *Journal of Marketing theory and Practice*, 19(2), 139-152. https://doi.org/10.2753/MTP1069-6679190202

Hair, J.F., Hult, G.T.M., Ringle, C.M. and Sarstedt, M. (2014), *A primer on Partial Least Squares Structural Equation Modeling (PLS-SEM)*, 2nd ed., Sage, Thousand Oaks.

Hayles, N. K. (2012). *How we think: Digital media and contemporary technogenesis*. University of Chicago Press.

Henseler, J., & Fassott, G. (2010). Testing moderating effects in PLS path models: An illustration of available procedures. *Handbook of partial least squares,* 713-735. Springer Berlin Heidelberg. https://doi.org/10.1007/978-3-540-32827-8_31

Hiltz, S. R., & Johnson, K. (1990). User satisfaction with computer-mediated communication systems. *Management science*, *36*(6), 739-764. https://doi.org/10.1287/mnsc.36.6.739

Hine, M. J., Murphy, S. A., Weber, M., & Kersten, G. (2009). The role of emotion and language in dyadic e-negotiations. *Group Decision and Negotiation*, 18(3), 193–211. https://doi.org/10.1007/s10726-008-9151-9

Holmbeck, G. N. (1997). Toward terminological, conceptual, and statistical clarity in the study of mediators and moderators: examples from the child-clinical and pediatric psychology literatures. *Journal of consulting and clinical psychology*, *65*(4), 599. https://doi.org/10.1037/0022-006X.65.4.599

Jain, A. K., & Dubes, R. C. (1988). *Algorithms for clustering data*. Prentice-Hall, Inc.

Jarke, M., Jelassi, M. T., & Shakun, M. F. (1987). Mediator: Towards a negotiation support system. *European Journal of Operational Research*, 31(3), 314–334. https://doi.org/10.1016/0377-2217(87)90041-5

Jensen, A. (2009). Discourse strategies in professional e-mail negotiation: A case study. *English for Specific Purposes*, 28(1), 4–18. https://doi.org/10.1016/j.esp.2008.10.002

Johnson, N. A., & Cooper, R. B. (2009). Media, affect, concession, and agreement in negotiation: IM versus telephone. *Decision Support Systems*, 46(3), 673–684. https://doi.org/10.1016/j.dss.2008.10.007

Kahan, J. P., & Katz, M. L. (1989). On the efficiency of scientific inquiry. *The RAND Journal of Economics*, 20(1), 3-22.

Kalyanaraman, S., & Sundar, S. S. (2006). The psychological appeal of personalized content in web portals: Does customization affect attitudes and behavior? *Journal of Communication*, 56(1), 110-132. https://doi.org/10.1111/j.1460-2466.2006.00006.x

Kerr, E. B., & Hiltz, S. R. (1982). *Computer-mediated communication systems: Status and evaluation.* Academic Press.

King, D. C., & Glidewell, J. C. (1980). Dyadic bargaining outcomes under individualistic and competitive orientations. *Human Relations*, 33(11), 781–803. https://doi.org/10.1177/001872678003301102

Koeszegi, S. T., Pesendorfer, E.-M., & Vetschera, R. (2011). Data-driven phase analysis of e-negotiations: An exemplary study of synchronous and asynchronous negotiations. *Group Decision and Negotiation*, 20(4), 385–410. https://doi.org/10.1007/s10726-008-9115-0

Kray, L. J., Galinsky, A. D., & Thompson, L. (2002). Reversing the gender gap in negotiations: An exploration of stereotype regeneration. *Organizational Behavior and Human Decision Processes,* 87(2), 386-409. https://doi.org/10.1006/obhd.2001.2979

Kurtzberg, T. R., Dunn-Jensen, L. M., & Matsibekker, C. L. Z. (2005). Multiparty e-negotiations: Agents, alliances, and negotiation success. *International Journal of Conflict Management,* 16(3), 245–264. https://doi.org/10.1108/eb022931

Kurtzberg, T. R., & Naquin, C. E. (2010). Electronic signatures and interpersonal trustworthiness in online negotiations. *Negotiation and Conflict Management Research,* 3(1), 49–63.
https://doi.org/10.1111/j.1750-4716.2009.00049.x

Lai, K. R., Lin, M. W., & Yu, T. J. (2010). Learning opponent's beliefs via fuzzy constraint-directed approach to make effective agent negotiation. *Applied Intelligence*, 33(2), 232–246. https://doi.org/10.1007/s10489-009-0162-2

Lakoff, G., & Johnson, M. (1980). *Metaphors we live by.* University of Chicago Press.

Lance, C. E. (1988). Residual centering, exploratory and confirmatory moderator analysis, and decomposition of effects in path models containing interactions. *Applied Psychological Measurement, 12*(2), 163-175.

Laubert, C., & Parlamis, J. (2019). Are you angry (happy, sad) or aren't You? Emotion detection difficulty in email negotiation. *Group Decision and Negotiation*, 28(2), 377–413. https://doi.org/10.1007/s10726-018-09611-4

Laux, D., Luse, A., & Mennecke, B. E. (2016). Collaboration, connectedness, and community: An examination of the factors influencing student persistence in virtual communities. *Computers in Human Behavior*, 57, 452-464.
https://doi.org/10.1016/j.chb.2015.12.046

Lewis, S. A., & Fry, W. R. (1977). Effects of visual access and orientation on the discovery of integrative bargaining alternatives. *Organizational Behavior and Human Performance*, 20(1), 75–92.
https://doi.org/10.1016/0030-5073(77)90045-9

Li, J., Li, W., & Ma, Y. (2020). Multi-channel negotiation mechanism for the buyer-seller problem based on clustering algorithm. *Journal of Intelligent & Fuzzy Systems*, 38(5), 5367-5380.

Liengaard, B. D., Sharma, P. N., Hult, G. T. M., Jensen, M. B., Sarstedt, M., Hair, J. F., & Ringle, C. M. (2021). Prediction: coveted, yet forsaken? Introducing a cross-validated predictive ability test in partial least squares path modeling. *Decision Sciences*, *52*(2), 362-392. https://doi.org/10.1111/deci.12445

Lim, J. (2000). An experimental investigation of the impact of NSS and proximity on negotiation outcomes. *Behaviour & Information Technology*, 19(5), 329–338. https://doi.org/10.1080/014492900750000036

Lim, X.-J., Cheah, J.-H., Waller, D.S., Ting, H. and Ng, S.I. (2019), "What s-commerce implies? Repurchase intention and its antecedents", *Marketing Intelligence & Planning*. https://doi.org/10.1108/MIP-03-2019-0145

Lim, J., & Yang, Y. P. (2008). Exploring computer-based multilingual negotiation support for English-Chinese dyads: can we negotiate in our native languages?. *Behaviour & Information Technology*, *27*(2), 139-151. https://doi.org/10.1080/01449290601111135

Little, T. D., Bovaird, J. A., & Widaman, K. F. (2006). On the merits of orthogonalizing powered and product terms: Implications for modeling interactions among latent variables. *Structural Equation Modeling*, *13*(4), 497-519. https://doi.org/10.1207/s15328007sem1304_1

Loewenstein, J., Thompson, L., & Gentner, D. (2003). Analogical learning in negotiation teams: Comparing cases promotes learning and transfer. *Academy of Management Learning & Education*, 2(2), 119-127. https://doi.org/10.5465/amle.2003.9901663

Loewenstein, J., Morris, M. W., Chakravarti, A., Thompson, L., & Kopelman, S. (2005). At a loss for words: Dominating the conversation and the outcome in negotiation as a function of intricate arguments and communication media. *Organizational Behavior and Human Decision Processes*, 98(1), 28–38. https://doi.org/10.1016/j.obhdp.2005.03.007

Marquandt, D. (1980). You should standardize the predictor variables in your regression models. Discussion of: A critique of some ridge regression methods. *Journal of the American Statistical Association*, *75*(369), 87-91.

McGinn, K. L., & Keros, A. T. (2002). Improvisation and the logic of exchange in socially embedded transactions. *Administrative Science Quarterly*, 47(3), 442–473. https://doi.org/10.2307/3094847

Moore, D. A., Kurtzberg, T. R., Thompson, L. L., & Morris, M. W. (1999). Long and short routes to success in electronically mediated negotiations: Group affiliations and good vibrations. *Organizational Behavior and Human Decision Processes*, 77(1), 22–43. https://doi.org/10.1006/obhd.1998.2814

Nenonen, S., Storbacka, K., & Windahl, C. (2010). Value creation in multichannel service contexts. *Journal of Service Management*, 21(1), 3-22.

Neslin, S. A., Grewal, D., Leghorn, R., Shankar, V., Teerling, M. L., Thomas, J. S., & Verhoef, P. C. (2006). Challenges and opportunities in multichannel customer management. *Journal of Service Research*, 9(2), 95–112. https://doi.org/10.1177/1094670506293559

Nyhart, J. D., & Samarasan, D. K. (1989). The elements of negotiation management: Using computers to help resolve conflict. *Negotiation Journal*, *5*(1), 43-62. https://doi.org/10.1007/BF01000811

Nunnally, J. C., & Bernstein, I. H. (1994). The assessment of reliability. *Psychometric theory*, *3*(1), 248-292.

Olekalns, M., & Smith, P. L. (2008). Restoring self-esteem in negotiation: An exploration of theory and practice. *Group Decision and Negotiation*, 17(2), 157-175.

Olekalns, M., & Smith, P. L. (2009). Behavioral negotiation theory: A framework for conceptualizing dyadic bargaining. *International Journal of Conflict Management*, 20(2), 150-171.

Paese, P. W., Schreiber, A. M., & Taylor, A. W. (2003). Caught telling the truth: Effects of honesty and communication media in distributive negotiations. *Group Decision and Negotiation*, *12*, 537-566. https://doi.org/10.1023/B:GRUP.0000004334.14310.90

Park, M. J., Choi, H., Rho, J. J. (2016). Citizen patronage behavior of government social media services: Extended perspective of perceived value with the moderating effect of media synchronicity. *Information Development*, 32(3), 293-312. https://doi.10.1177/0266666914543959

Park, M. J., Choi, H., Rho, J. J. (2016). Citizen patronage behavior of government social media services: Extended perspective of perceived value with the moderating effect of media synchronicity. *Information Development*, 32(3), 293-312. https://doi.10.1177/0266666914543959

Payne, A. F., Storbacka, K., & Frow, P. (2008). Managing the co-creation of value. *Journal of the Academy of Marketing Science*, 36(1), 83-96. https://doi.org/10.1007/s11747-007-0070-0

Pedhazur, E. J., & Kerlinger, F. N. (1982). *Multiple regression in behavioral research*. Holt, Rinehart, and Winston.

Pesendorfer, E.-M., Graf, A., & Koeszegi, S. T. (2007). Relationship in electronic negotiations: Tracking behavior over time. *Journal of Business Economics*, 77(12), 1315–1338. https://doi.org/10.1007/s11573-007-0327-6

Pesendorfer, E.-M., & Koeszegi, S. T. (2006). Hot versus cool behavioural styles in electronic negotiations: The impact of communication mode. *Group Decision and Negotiation*, 15(2), 141–155. https://doi.org/10.1007/s10726-006-9025-y

Plato. *Plato in Twelve Volumes*, Vol. 1 translated by Harold North Fowler; Introduction by W.R.M. Lamb. Cambridge, MA, Harvard University Press; London, William Heinemann Ltd. 1966.

Robert, L. P., & Dennis, A. R. (2005). Paradox of richness: A cognitive model of media choice. *IEEE Transactions on Professional Communication*, 48(1), 10–21. https://doi.org/10.1109/TPC.2004.843292

Scheck, S., Allmendinger, K., & Hamann, K. (2008). The effects of media richness on multilateral negotiations in a collaborative virtual environment. *Journal of Media Psychology*, 20(2), 57–66. https://doi.org/10.1027/1864-1105.20.2.57

Schoop, M., Köhne, F., Staskiewicz, D., Voeth, M., & Herbst, U. (2008). The antecedents of renegotiations in practice—An exploratory analysis. *Group Decision and Negotiation*, 17(2), 127–139. https://doi.org/10.1007/s10726-007-9080-z

Schulz, J. W., & Pruitt, D. G. (1978). The effects of mutual concern on joint welfare. *Journal of Experimental Social Psychology*, 14(5), 480–492. https://doi.org/10.1016/0022-1031(78)90044-6

Segars, A. H., & Grover, V. (1993). Re-examining perceived ease of use and usefulness: A confirmatory factor analysis. *MIS quarterly*, 517-525. https://doi.org/10.2307/249590

Sharma, P. N., Liengaard, B. D., Hair, J. F., Sarstedt, M., & Ringle, C. M. (2022). Predictive model assessment and selection in composite-based modeling using PLS-SEM: Extensions and guidelines for using CVPAT. *European Journal of Marketing*. https://doi.org/10.1108/EJM-08-2020-0636

Sharma, P., Sarstedt, M., Shmueli, G., Kim, K. H., & Thiele, K. O. (2019). PLS-based model selection: The role of alternative explanations in information systems research. *Journal of the Association for Information Systems*, *20*(4), 4. https://doi.10.17005/1.jais.00538

Sheffield, J. (1995). The effect of communication medium on negotiation performance. *Group Decision and Negotiation*, 4(2), 159–179. https://doi.org/10.1007/BF01410100

Shmueli, G., Sarstedt, M., Hair, J. F., Cheah, J. H., Ting, H., Vaithilingam, S., & Ringle, C. M. (2019). Predictive model assessment in PLS-SEM: guidelines for using PLSpredict. *European journal of marketing, 53*(11), 2322-2347. https://doi.org/10.1108/EJM-02-2019-0189

Smith, D. H. (1969). Communication and negotiation outcome. *Journal of Communication*, 19(3), 248–256. https://doi.org/10.1111/j.1460-2466.1969.tb00847.x

Smith, K. W., & Sasaki, M. S. (1979). Decreasing multicollinearity, a method for models with multiplicative functions. *Sociological Methods & Research, 8*(1), 35-56.

Spangle, & Isenhart. (2002). *Negotiation: Communication for diverse settings*. SAGE Publications.

Spiller, S. A., Fitzsimons, G. J., Lynch Jr, J. G., & McClelland, G. H. (2013). Spotlights, floodlights, and the magic number zero: Simple effects tests in moderated regression. *Journal of Marketing Research*, *50*(2), 277-288.

Tate, R. L. (1984). Limitations of centering for interactive models. *Sociological Methods & Research, 13*(2), 251-271.

Wedel, M., & Kamakura, W. A. (2002). Introduction to the special issue on market segmentation. *Intern. J. of Research in Marketing, 19*, 181-183. https://ssrn.com/abstract=2395277

Thompson, L., Wang, J., & Gunia, B. (2010). Negotiation. *Annual Review of Psychology*, 61, 491-515. https://doi.org/10.1146/annurev.psych.093008.100458

Ting, H., de Run, E.C., Cheah, J.-H. and Chuah, F. (2016), Food neophobia and ethnic food consumption intention: An extension of the theory of planned behaviour, *British Food Journal*, Vol. 118 No. 11, 2781-2797. https://doi.org/10.1108/BFJ-12-2015-0492

Townley, A., & Jones, A. (2016). The role of emails and covering letters in negotiating a legal contract: A case study from Turkey. *English for Specific Purposes*, 44, 68–81. https://doi.org/10.1016/j.esp.2016.07.001

Van de Ven, A. H. (1976). On the nature, formation, and maintenance of relations among organizations. *Academy of management review*, 1(4), 24-36. https://doi.org/10.2307/2094477

Van de Ven, A. H., Delbecq, A. L., & Koenig Jr, R. (1976). Determinants of co-ordination modes within organizations. *American sociological review*, 322-338. https://doi.org/10.2307/2094477

Vinciarelli, A., Salamin, H., & Polychroniou, A. (2014). Negotiating over mobile phones: Calling or being called can make the difference. *Cognitive Computation*, 6(4), 677–688. https://doi.org/10.1007/s12559-014-9267-y

Volkema, R., Fleck, D., & Hofmeister, A. (2010). Predicting competitive-unethical negotiating behavior and its consequences: Competitive-unethical negotiating behavior. *Negotiation Journal*, 26(3), 263–286. https://doi.org/10.1111/j.1571-9979.2010.00273.x

Wang, Y. M., Lin, H. H., Tai, W. C., & Fan, Y. L. (2016). Understanding multi-channel research shoppers: an analysis of Internet and physical channels. *Information Systems and e-Business Management*, 14, 389-413. https://doi.org/10.1007/s10257-015-0288-1

Wang, W. Y., Pauleen, D. J., & Zhang, T. (2016). How social media applications affect B2B communication and improve business performance in SMEs. *Industrial Marketing Management*, 54, 4-14. https://doi.org/10.1016/j.indmarman.2015.12.004

Wang, X., & Liu, Y. (2021). Multi-channel negotiation decision-making based on clustering algorithm in e-commerce. *Cluster Computing*, 24(2), 1181-1191.

Weeks, G. D., & Chapanis, A. (1976). Cooperative versus conflictive problem solving in three telecommunication modes. *Perceptual and Motor Skills*, 42(3), 879–917. https://doi.org/10.2466/pms.1976.42.3.879

Weingart, L. R., Todorova, G., & Prietula, M. J. (2015). Comments on "culture and negotiation strategy". *Negotiation and Conflict Management Research*, 8(2), 101-106.

Wolfe, C. J., & Murthy, U. S. (2005). Negotiation support systems in budget negotiations: An experimental analysis. *Journal of Management Information Systems*, 22(3), 351–381. https://doi.org/10.2753/MIS0742-1222220312

Woodward, J. (1965). *Industrial Organisation: Theory and Practice* (Oxford, U.P.).

Xie, B., Zhan, J., & Wang, X. (2020). A clustering-based approach to multi-channel negotiation strategy in electronic commerce. *Applied Sciences*, 10(10), 3459.

Appendices

Annex 1. Online Questionnaire on Multi-Channel Negotiations

Table 1. Online Questionnaire on Multi-Channel Negotiations 1/3

Item	Topic	Question / Content	Answer Categories
(1)	Participation	Did you participate in a negotiation in the last 12 months?	☐ Yes ☐ No
(2)	Exit-Filter	As you did not participate in a negotiation in the past 12 months, the survey ends here.	
(3)	Channels	Did you use at least two negotiation channels during your last relevant negotiation? It is important that your last relevant negotiation is already closed and completed. A possible negotiation channel could be face-to-face, a phone conversation, email, chat, messaging, an audio conference, or video conference.	☐ Yes ☐ No
(4)	Choice	Which negotiation channel or channels did you use in your last significant negotiation? Please think of one significant negotiation that you conducted or were part of in the past 12 months. It is important that your last relevant negotiation is already closed and completed. When thinking of this significant negotiation, kindly select which negotiation channel or channels you used. You can select one or multiple negotiation channels if needed.	☐ Phone Call ☐ Face-to-Face ☐ Email ☐ Chat ☐ Messaging ☐ Audio Conferencing ☐ Video Conferencing ☐ Other
(5)	Percentage Share	If 100% is your total negotiation channel usage, how much in percentage share did you use the following channels? As an example, you could indicate face-to-face 30%, email 20%, video 30%, messaging 20%. For the percentage share please only type in whole numbers (e.g., typing-in a numerical of 30 is acceptable). Please do not type in decimal numbers as these are not acceptable (e.g.,0.25 is not possible).	☐ Phone Call ☐ Face-to-Face ☐ Email ☐ Chat ☐ Messaging ☐ Audio Conferencing ☐ Video Conferencing

Table 2. Online Questionnaire on Multi-Channel Negotiations 2/3

Item	Topic	Question / Content	Answer Categories
(6)	Preference	Do you have a personal preference for a particular negotiation channel in general when you negotiate? If yes, kindly select the preferred negotiation channel from the list below.	☐ Phone Call ☐ Face-to-Face ☐ Email ☐ Chat ☐ Messaging ☐ Audio Conferencing ☐ Video Conferencing ☐ Other
(7)	Purpose	In general, do you select specific negotiation channels on purpose for your negotiation?	☐ Yes ☐ No

Table 3. Online Questionnaire on Multi-Channel Negotiations 3/3

Item	Negotiation Stage	Question / Content	Answer Categories
(8)	Greeting & Initiation (1)	A negotiation can consist of various stages. Please think of your last significant negotiation. It is important that your last relevant negotiation is already closed and completed. If you applied this negotiation stage, kindly indicate which channel or channels you used.	☐ Phone Call ☐ Face-to-Face ☐ Email ☐ Chat ☐ Messaging ☐ Audio Conferencing ☐ Video Conferencing ☐ Other ☐ This negotiation stage was not required
(9)	Background Information Exchange (2)		
(10)	Agenda Setting (3)		
(11)	Issue Discussion (4)		
(12)	Final Agreement Formulation (5)		

Table 4. Latent Constructs 1/2

While thinking of your last significant negotiation, kindly evaluate the following statements on a scale from 1 (strongly disagree) to 5 (strongly agree).

(13) Self Efficacy	Strongly disagree				I strongly agree
	1	2	3	4	5
I feel competent enough to select the best negotiation channels and channel mix for the negotiation.					
I think that I have the relevant knowledge and expertise to make solid evaluations regarding the choice of channels for the negotiation.					
I had difficulties evaluating the suitability of the channels and the channel mix for the negotiation.					
I can choose channels to a certain degree; however operating procedures determine the channels to use.					

(14) Agility	Strongly disagree				Strongly agree
	1	2	3	4	5
Multi-channel negotiations enable us to make decisions quickly when circumstances in the negotiation change.					
Decisions in multi-channel negotiations are taken and implemented very quickly.					
Information in multi-channel negotiations is handled in real-time.					
Multi-channel negotiations allow resources (material, financial, human) to be easily deployed to respond to opportunities and threats encountered during the negotiation.					

(15) Synchronicity 1	Strongly disagree				Strongly agree
	1	2	3	4	5
I am more inclined to use multiple negotiation channels to enhance the negotiation relationship to gain more information in appropriate time.					
A multi-channel negotiation is the appropriate media for creating a shared understanding among negotiators.					
Multi-channel negotiations fit well with the demand to deliver information.					

Table 5. Testing Synchronicity Theory

Item	Topic	Question / Content	Answer Categories
(16)	Synchronicity 2	Which is the most important channel to you for sharing information with the negotiation party? Please rank in order of importance. With information sharing we refer to the exchange of facts and background information that are helpful to the course of the negotiation.	☐ Phone Call ☐ Face-to-Face ☐ Email ☐ Chat ☐ Messaging ☐ Audio Conferencing ☐ Video Conferencing
(17)	Synchronicity 3	Which is the most important negotiation channel you use to reach a consensus or decision? Please rank in order of importance.	☐ Phone Call ☐ Face-to-Face ☐ Email ☐ Chat ☐ Messaging ☐ Audio Conferencing ☐ Video Conferencing

Tables 6-11. Latent Constructs 2/2

While thinking of your last significant negotiation, kindly evaluate the following statements on a scale from 1 (strongly disagree) to 5 (strongly agree).

(18) Behavioral Control	Strongly disagree				Strongly agree
	1	2	3	4	5
I feel in control of the negotiation using multiple negotiation channels.					
Using multiple negotiation channels in the negotiation gives an opportunity to the negotiation party to be in charge.					
While using multiple negotiation channels in my negotiation, I feel decisive.					
Multi-channel negotiations give me more control over the negotiation outcome.					

(19) Complexity	Strongly disagree				Strongly agree
	1	2	3	4	5
The negotiation consisted of many negotiation channels.					
The use of various negotiation channels is very different from using only one channel.					
The use of various negotiation channels during the negotiation requires a different input of time depending on the channel.					
Negotiation channels are individually selected for this specific negotiation.					
The use of many negotiation channels requires intense alignments with involved parties.					
The number of negotiation channels changes often.					
The diversity of negotiation channels changes often.					

(20) Performance Risk	Strongly disagree				Strongly agree
	1	2	3	4	5
There is a high chance that there could be something wrong with a multi-channel negotiation or that the multi-channel negotiation results are not delivered as promised.					
There is a high chance of loss in the negotiation because multiple channels are not used well.					
A negotiation with multiple channels is extremely risky in terms of how it performs.					

(21) Cognitive Resource Demands	Not a lot of though				A lot of thought
	1	2	3	4	5
While thinking of your last significant single or multi-channel negotiation, kindly evaluate if your negotiation required on a scale from 1 (not a lot of thought) to 5 (a lot of thought).					
	Not a lot of focus				A lot of focus
	1	2	3	4	5
While thinking of your last significant single or multi-channel negotiation, kindly evaluate if your negotiation required on a scale from 1 (not a lot of focus) to 5 (a lot of focus).					
	Not a lot of concentration				A lot of concentration
	1	2	3	4	5
While thinking of your last significant single or multi-channel negotiation, kindly evaluate if your negotiation required on a scale from 1 (not a lot of concentration) to 5 (a lot of concentration).					

(22) Negotiation Effectiveness	Strongly disagree				Strongly agree
	1	2	3	4	5
During the negotiation, we were able to achieve an optimal outcome regarding the negotiation objects.					
The goals defined for the negotiation process were fully achieved.					
Examining the current negotiation, the attained negotiation outcome is optimal.					
The achieved goal has a positive impact on future relationships with our negotiation partners.					
In the current negotiation situation, a better outcome would realistically not be possible.					

(23) Negotiation Efficiency	Strongly disagree				Strongly agree
	1	2	3	4	5
During the negotiation, we only focused on relevant aspects.					
All negotiation steps of the negotiation were necessary to achieve the negotiation target.					
Overall, the negotiation was very goal oriented and structured.					
A negotiation goal was achieved fast.					
The negotiation goal was achieved by involving the least cost / expenses.					

Table 12. Negotiation Experience and Negotiation History

Item	Topic	Question / Content	Answer Categories
(24)	Negotiation Experience	Do you have experience in negotiations?	☐ Yes ☐ No
(25)	Negotiation History	In your last significant negotiation, did you have a prior negotiation history with the negotiation party or parties? Kindly type-in the duration of your negotiation history with the negotiation party in years. The negotiation history must be at least one year. You can only type-in numerical values, text is not possible.	☐ Yes ☐ No

Table 13. Demographics 1/2

Item	Topic	Question / Content	Answer Categories
(26)	Demographics 1	How old are you? You can only type-in numerical values, text is not possible.	☐ Kindly let us know your age in the box below. ☐ I prefer not to disclose my age.
		Which gender are you?	☐ Female ☐ Male ☐ Diverse ☐ I prefer not to disclose my gender.
		What is your nationality?	☐ Kindly type your nationality in the box below. ☐ I prefer not to disclose my nationality.

Table 14. Demographics 2/2

Item	Topic	Question / Content	Answer Categories
(27)	Demographics 2	In which functional area are you operating in your organization?	☐ Management ☐ Purchasing ☐ Sales ☐ Project Management ☐ Marketing ☐ Business Development ☐ M&A / Finance ☐ Other
		What is your current position?	☐ Administrator, Team Member, Assistant ☐ Manager without a team ☐ Team Lead, Group Lead ☐ Department Head ☐ Division Manager, Area Manager, Business Unit Manager, Senior Department Head ☐ Managing Director, Board Member, Executive ☐ Other
		In which industry do you work?	☐ Automotive ☐ Consumer Goods ☐ Media ☐ Telecommunications ☐ Chemicals ☐ Retail ☐ Real Estate ☐ Petrochemical Sector / Oil Industry ☐ Pharmaceuticals ☐ Industrial Goods ☐ High Technology Industry ☐ Public Sector ☐ Other
		How many employees work in your organization?	☐ < 50 ☐ 50 - 250 ☐ 251 - 1000 ☐ 1001 - 5000 ☐ 5001 - 15000 ☐ 15000 - 50000 ☐ > 50000
		In which country do you work?	☐ Kindly insert the country where you work in the box below. ☐ I prefer not to disclose the location of my work.

Table 15. Closing of Survey

Item	Topic	Question / Content
(28)	Email Study Results	If you would like to receive the results of this study, kindly type your email address into the box below.
(29)	End of Survey	Thank you for your participation in this study.

Annex 2. Questionnaire Results for Latent Variables

Figure 1. Answering Scale Item 1 for Self-Efficacy

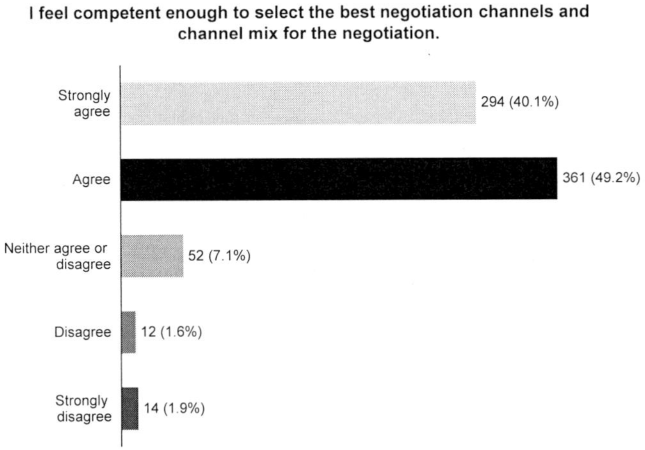

Note: n = 733

Figure 2. Answering Scale Item 2 for Self-Efficacy

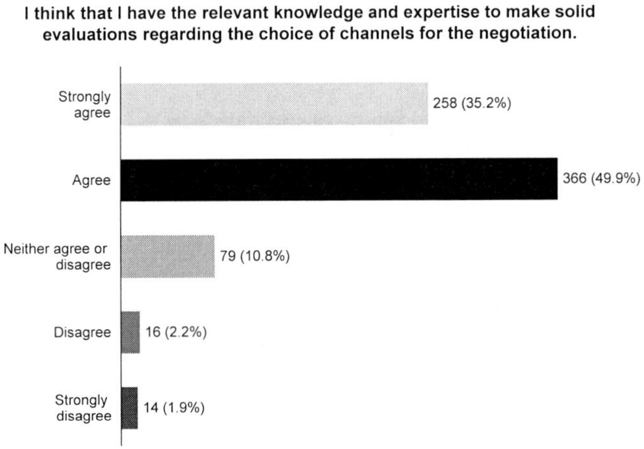

Note: n = 733

Figure 3. Answering Scale Item 3 for Self-Efficacy

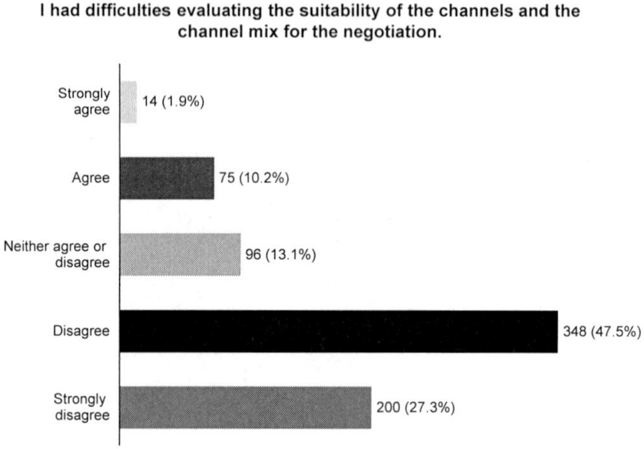

Note: n = 733

Figure 4. Answering Scale Item 1 for Agility

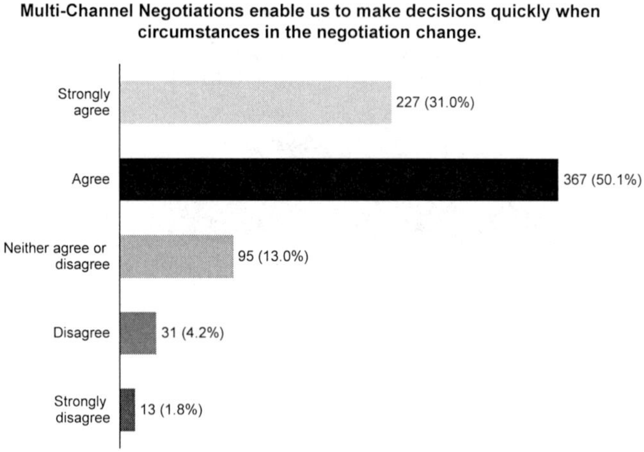

Note: n = 733

Figure 5. Answering Scale Item 2 for Agility

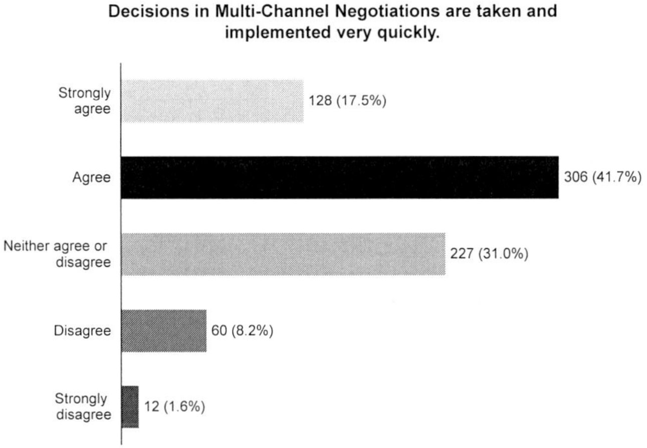

Note: n = 733

Figure 6. Answering Scale Item 3 for Agility

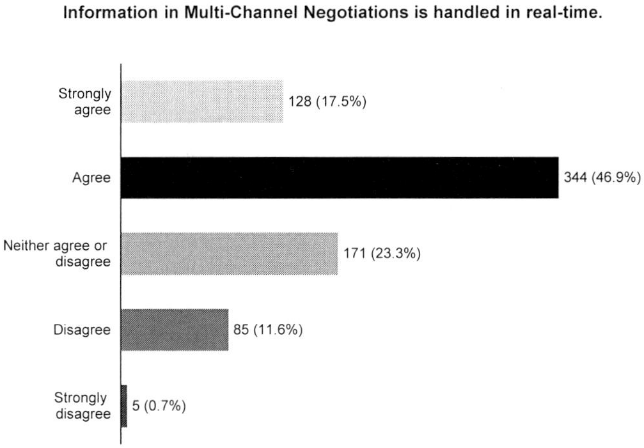

Note: n = 733

Figure 7. Answering Scale Item 4 for Agility

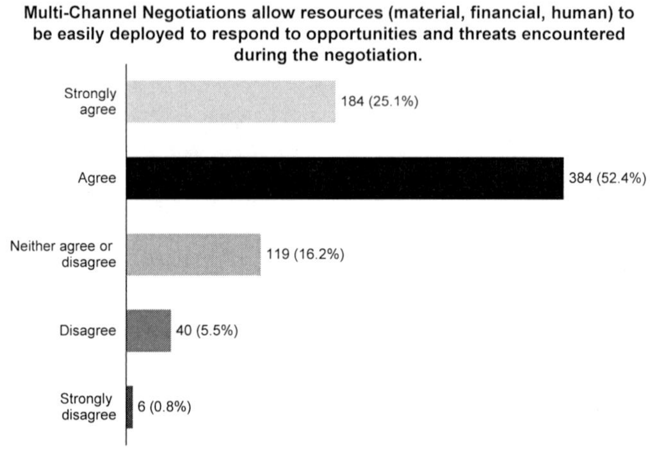

Note: n = 733

Figure 8. Answering Scale Item 1 for Synchronicity

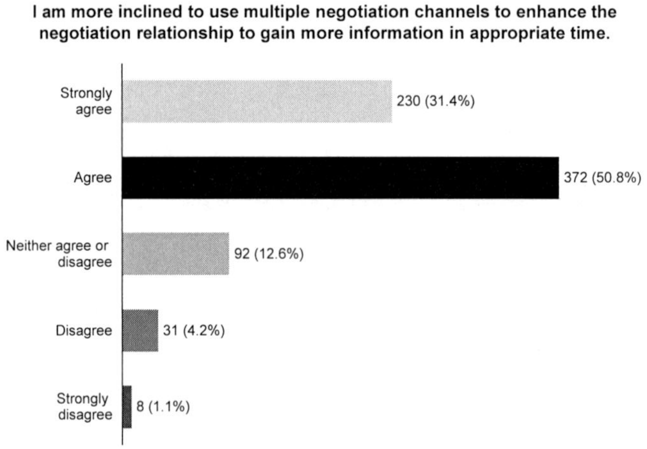

Note: n = 733

Figure 9. Answering Scale Item 2 for Synchronicity

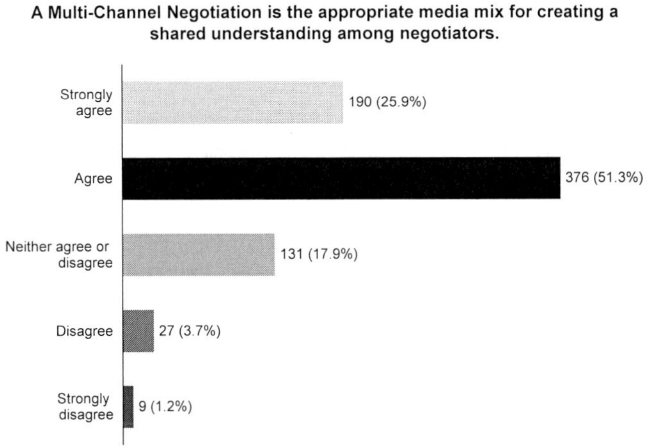

Note: n = 733

Figure 10. Answering Scale Item 3 for Synchronicity

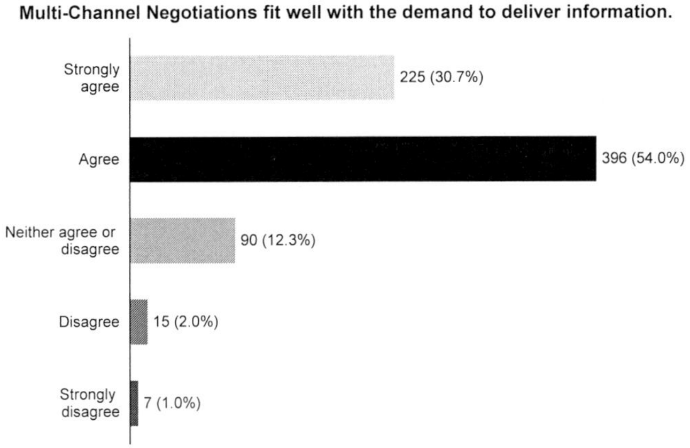

Note: n = 733

Figure 11. Answering Scale Item 1 for Behavioral Control

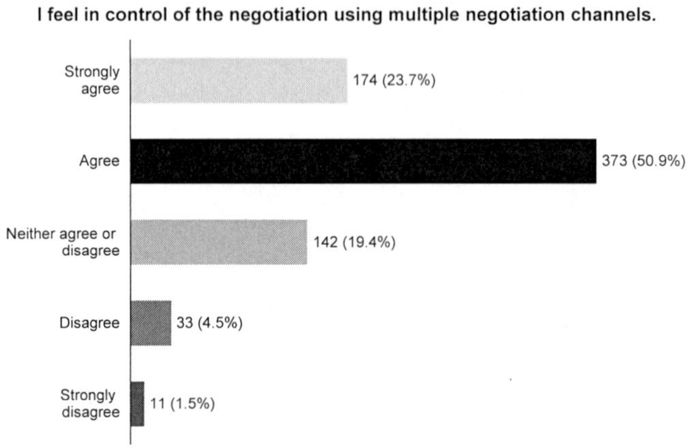

Note: n = 733

Figure 12. Answering Scale Item 2 for Behavioral Control

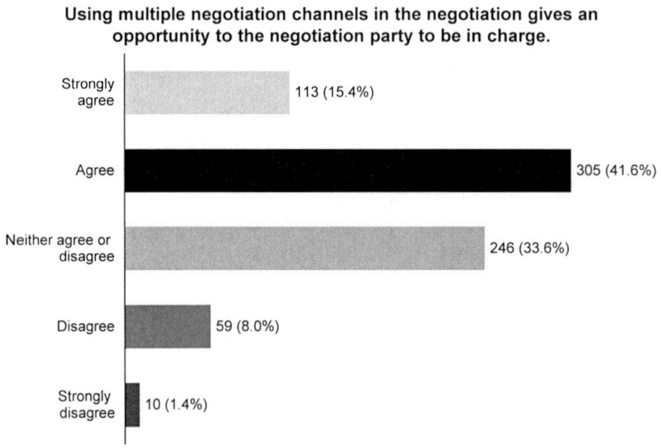

Note: n = 733

Figure 13. Answering Scale Item 3 for Behavioral Control

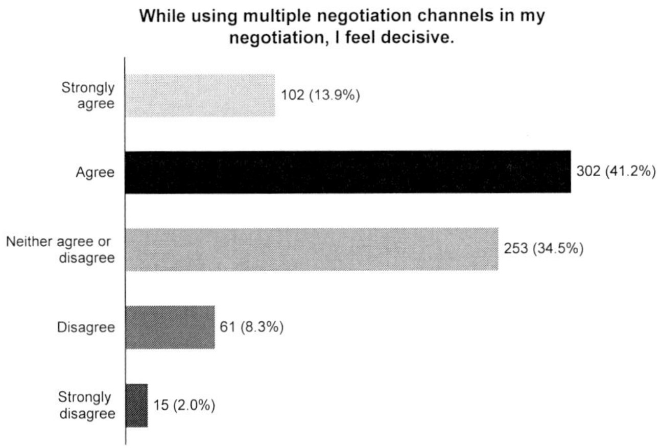

Note: n = 733

Figure 14. Answering Scale Item 4 for Behavioral Control

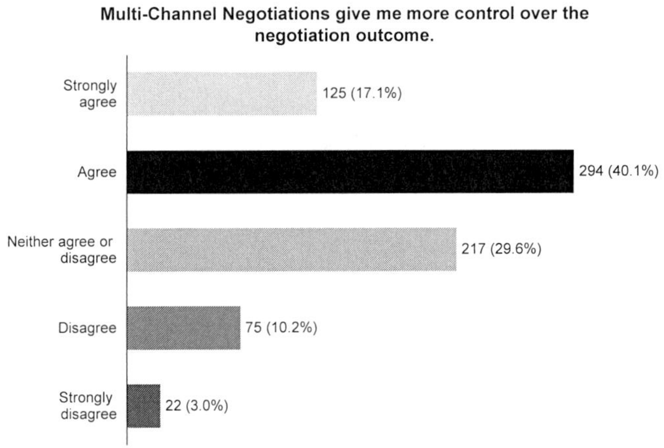

Note: n = 733

Figure 15. Answering Scale Item 1 for Complexity

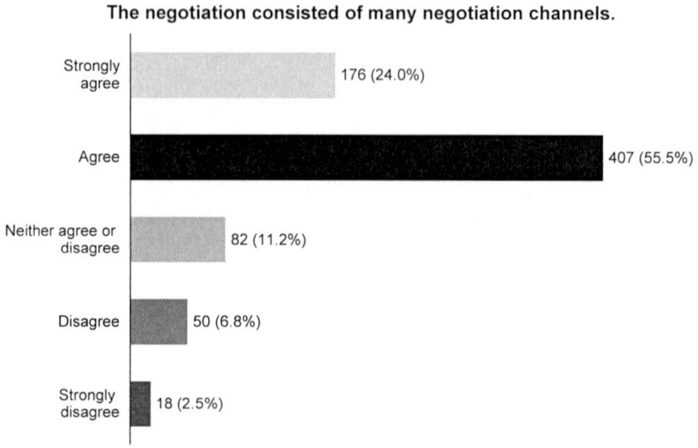

Note: n = 733

Figure 16. Answering Scale Item 2 for Complexity

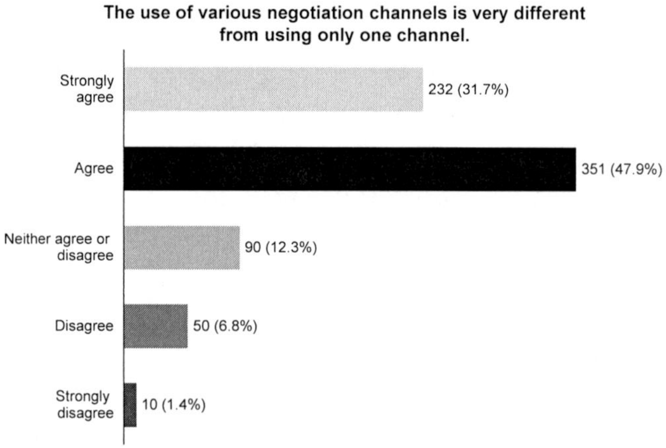

Note: n = 733

Figure 17. Answering Scale Item 3 for Complexity

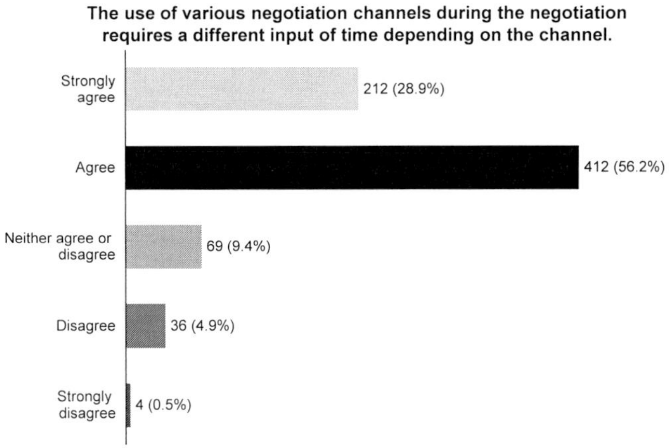

Note: n = 733

Figure 18. Answering Scale Item 4 for Complexity

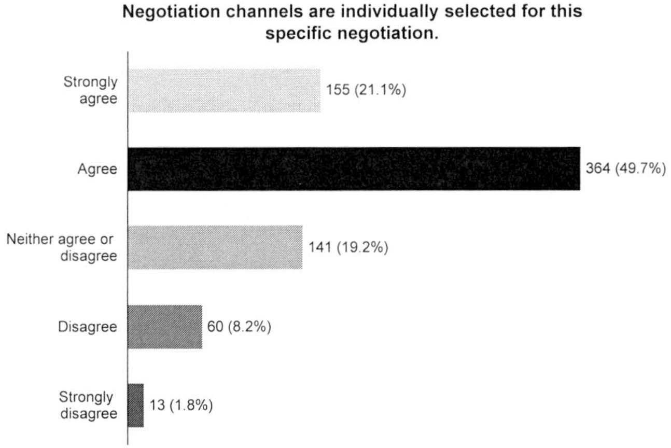

Note: n = 733

Figure 19. Answering Scale Item 5 for Complexity

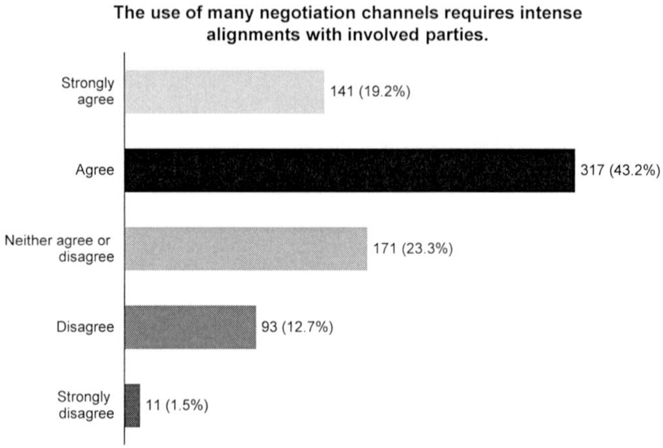

Note: n = 733

Figure 20. Answering Scale Item 6 for Complexity

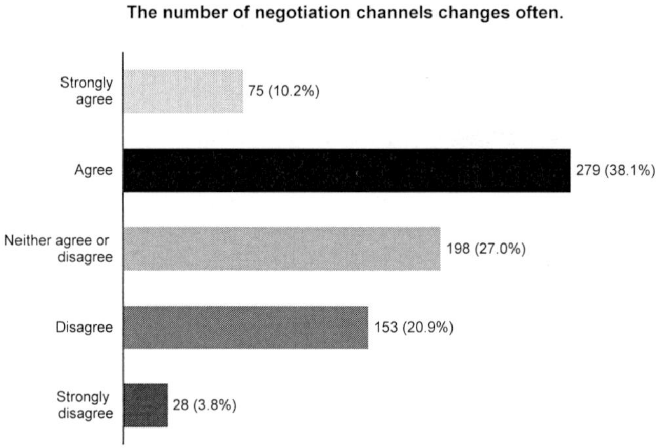

Note: n = 733

Figure 21. Answering Scale Item 7 for Complexity

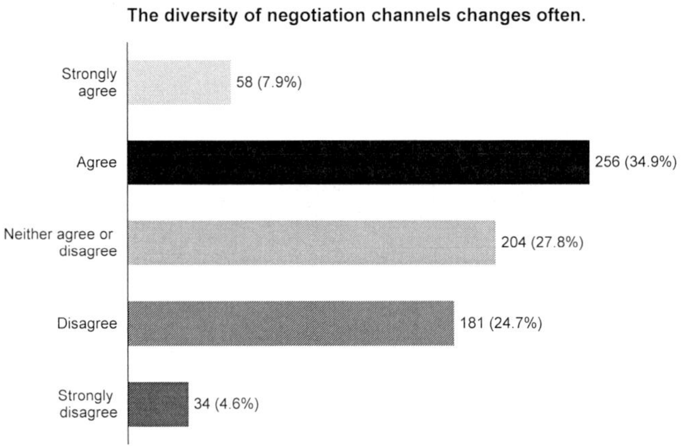

Note: n = 733

Figure 22. Answering Scale Item 1 for Performance Risk

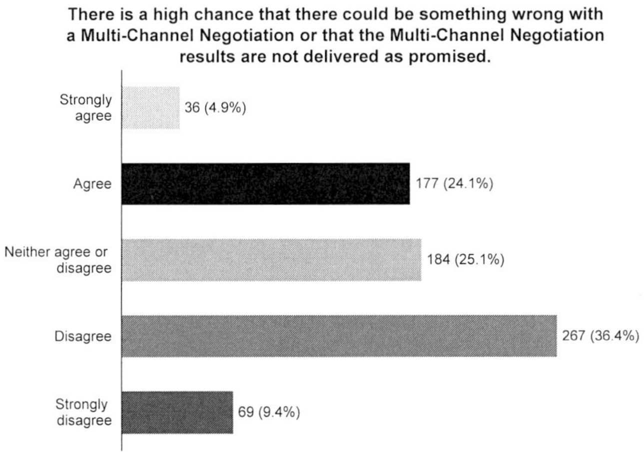

Note: n = 733

Figure 23. Answering Scale Item 2 for Performance Risk

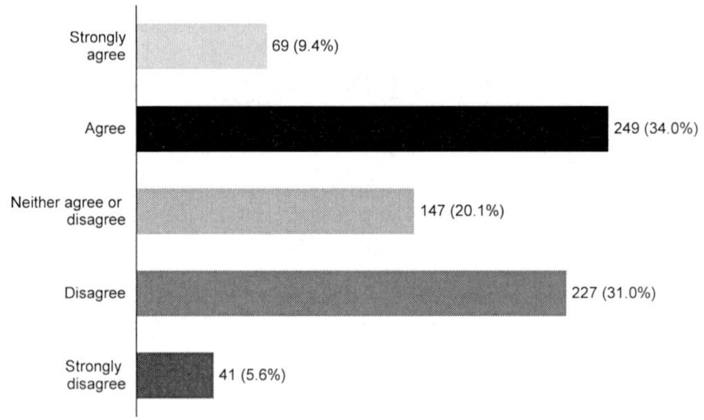

Note: n = 733

Figure 24. Answering Scale Item 3 for Performance Risk

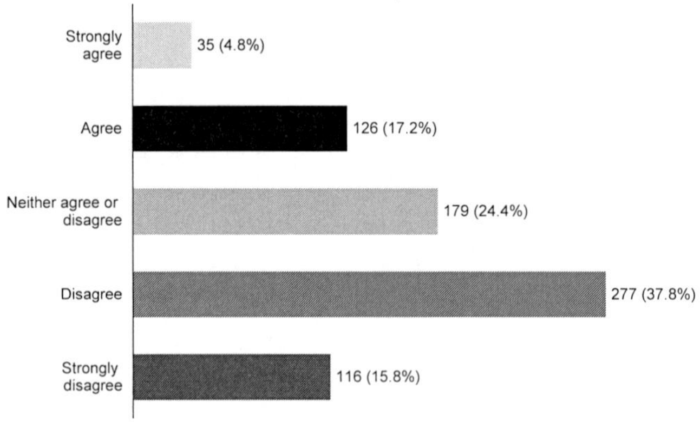

Note: n = 733

Figure 25. Answering Scale Item 1 for Cognitive Resource Demands

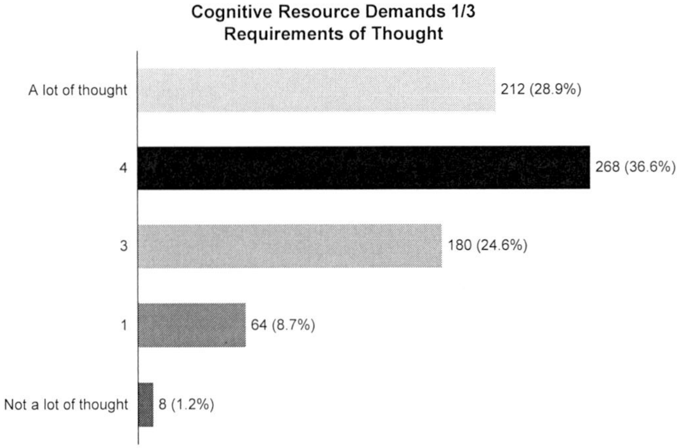

Note: n = 733

Figure 26. Answering Scale Item 2 for Cognitive Resource Demands

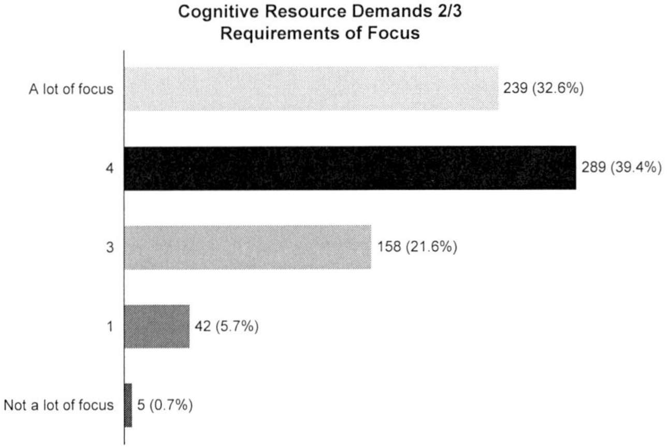

Note: n = 733

Figure 27. Answering Scale Item 3 for Cognitive Resource Demands

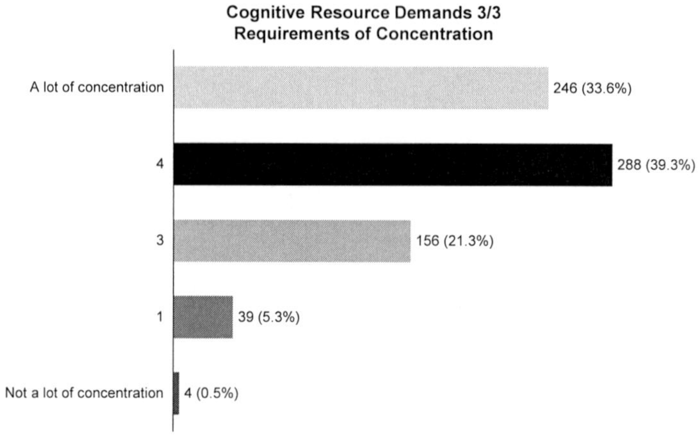

Note: n = 733

Figure 28. Answering Scale Item 1 for Negotiation Effectiveness

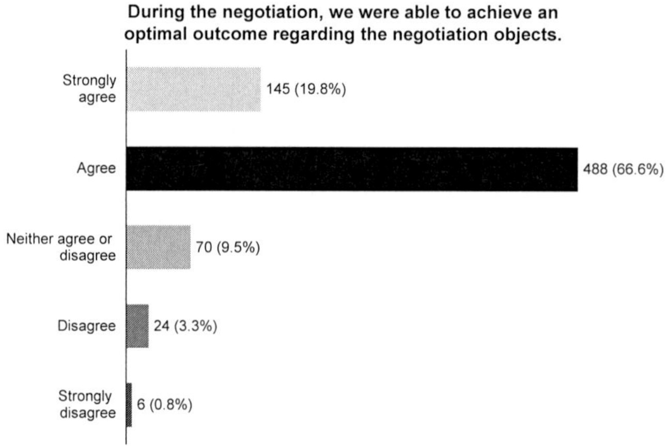

Note: n = 733

Figure 29. Answering Scale Item 2 for Negotiation Effectiveness

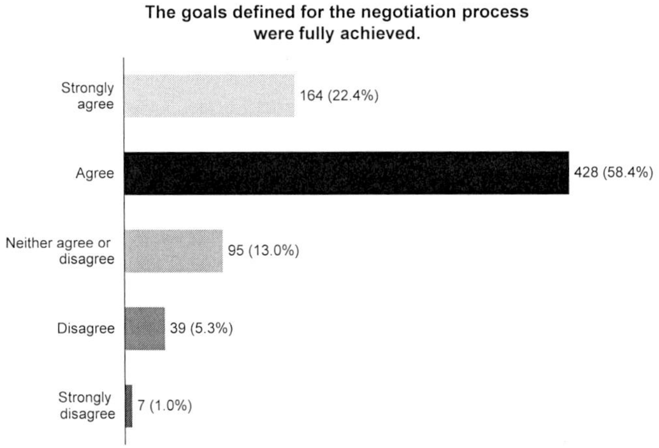

Note: n = 733

Figure 30. Answering Scale Item 3 for Negotiation Effectiveness

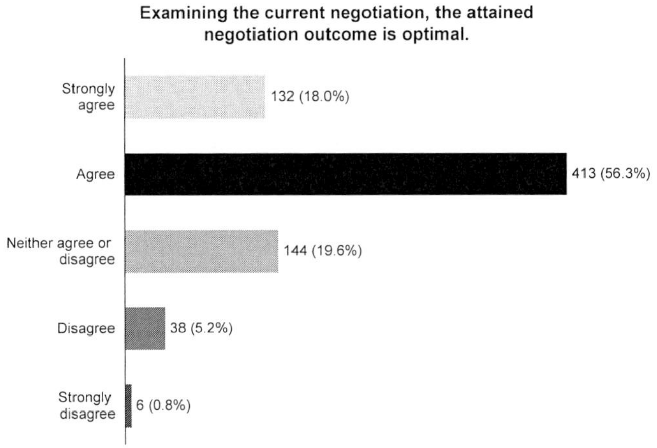

Note: n = 733

Figure 31. Answering Scale Item 4 for Negotiation Effectiveness

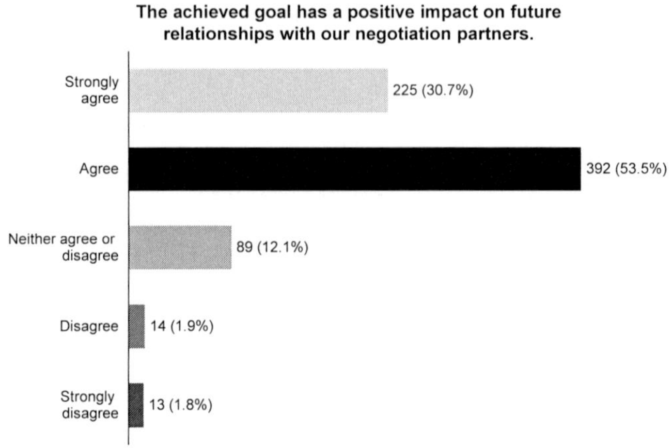

Note: n = 733

Figure 32. Answering Scale Item 5 for Negotiation Effectiveness

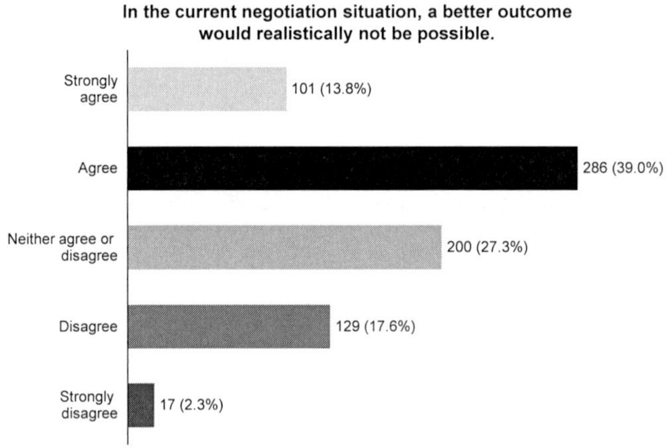

Note: n = 733

Figure 33. Answering Scale Item 1 for Negotiation Efficiency

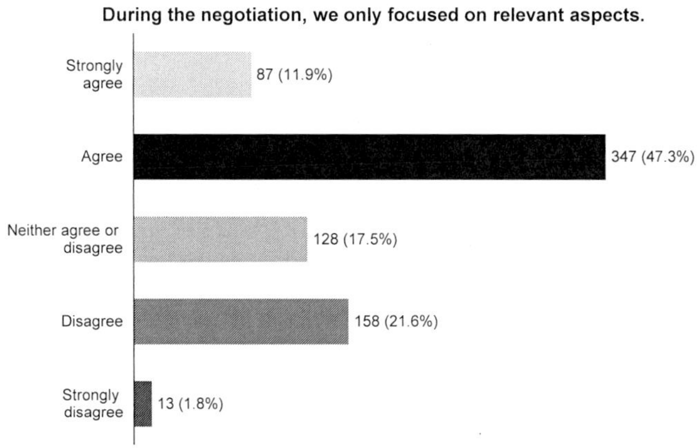

Note: n = 733

Figure 34. Answering Scale Item 2 for Negotiation Efficiency

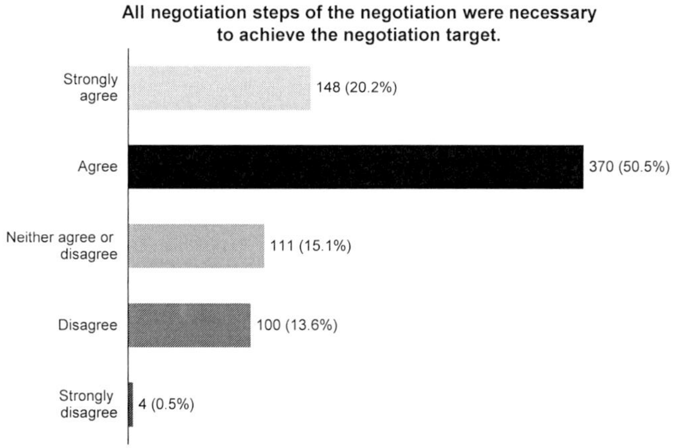

Note: n = 733

Figure 35. Answering Scale Item 3 for Negotiation Efficiency

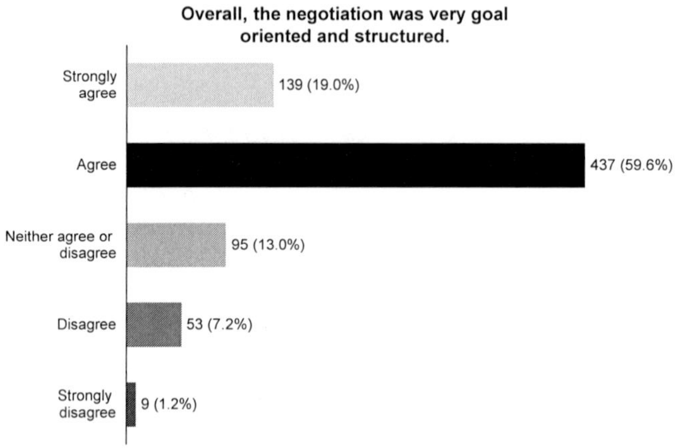

Note: n = 733

Figure 36. Answering Scale Item 4 for Negotiation Efficiency

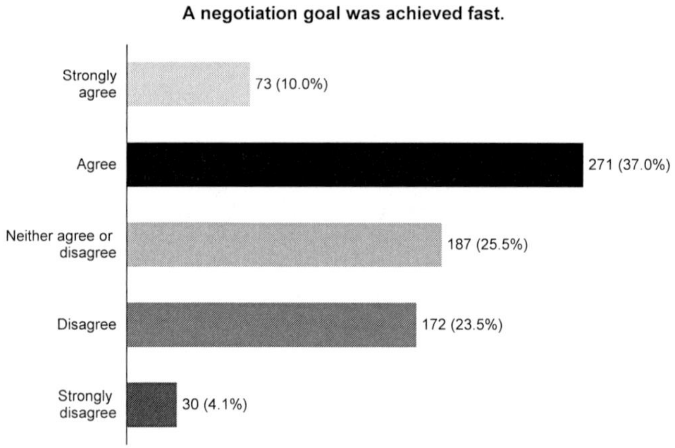

Note: n = 733

Figure 37. Answering Scale Item 5 for Negotiation Efficiency

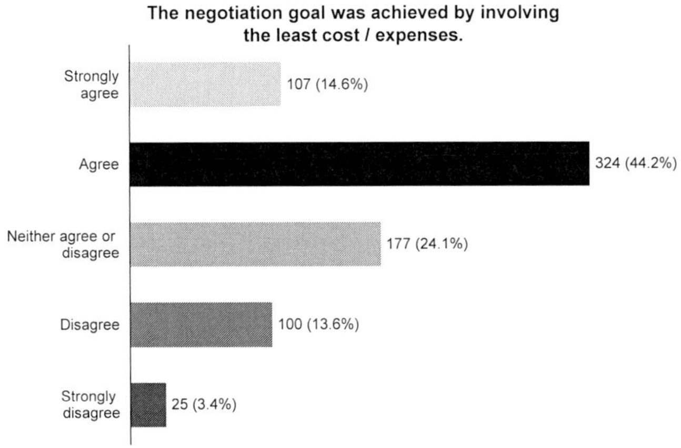

Note: n = 733

Annex 3. Measurement Scales for Behavioral Dimensions

Table 1. Self-Efficacy Measure

Self-Efficacy
(1) Multi-channel negotiations enable us to make decisions quickly when circumstances in the negotiation change.
(2) Decisions in multi-channel negotiations are taken and implemented very quickly.
(3) Information in multi-channel negotiations is handled in real time (reverse item).*
~~(4) I can choose channels to a certain degree; however operating procedures determine the channels to use.~~

Source: Adapted from Fuchs et al., 2010; Bandura, 1977; McAuley & Mihalko, 1998; DeSanctis & Poole, 1994; Note: *Item (3) is a reverse item and Item (4) is deleted for achievement of internal consistency.

Table 2. Agility Measure

Agility
(1) Multi-channel negotiations enable us to make decisions quickly when circumstances in the negotiation change.
(2) Decisions in multi-channel negotiations are taken and implemented very quickly.
(3) Information in multi-channel negotiations is handled in real time.
(4) Our resources (material, financial, human) are easily deployed via multiple negotiation channels to respond to opportunities and threats encountered during the negotiation.

Source: Adapted from Charbonnier-Voirin, 2011

Table 3. Synchronicity Measure

Synchronicity
(1) I am more inclined to use multiple negotiation channels to enhance the negotiation relationships to gain more information in appropriate time.
(2) With regard to media capabilities, multi-channel negotiation is the appropriate media mix for a convergence process focusing on a shared understanding among negotiators.
(3) The characteristics of multi-channel negotiations fit well with the features of the communication process and information transaction in negotiations.

Source: Adapted from Dennis et al, 2008, 2009; Ryoo & Koo, 2010; Park et al., 2014

Table 4. Behavioral Control Measure

Behavioral Control
(1) I feel in control of the negotiations using multiple negotiation channels.
(2) Using multiple negotiation channels in the negotiation gives an opportunity to the negotiation party to be in charge.
(4) While using multiple negotiation channels in my negotiation, I feel decisive.
(3) Multi-channel negotiations give me more control over the negotiation outcome.

Source: Adapted from Collier & Sherrell, 2010; Rippé et al., 2017

Annex 4. Measurement Scales for the Impact of Moderating Effects on the Process and Outcome in Multi-Channel Negotiations

Table 1. Measurement for the Moderating Effect Complexity

Complexity
(1) The negotiation consisted of many negotiation channels.
(2) The use of various negotiation channels is very different from using only one channel.
(5) The use of various negotiation channels during the negotiation requires a different input of time depending on the channel.
(6) Negotiation channels are individually selected for this specific negotiation.
(7) The use of many negotiation channels requires intense alignments with involved parties.
(8) The number of negotiation channels changes often.
(9) The diversity of negotiation channels changes often.

Source: Adapted from Braun & Hadwich, 2016

Table 2. Measurement for the Moderator Cognitive Resource Demands

Cognitive Resource Demands
(1) While thinking of your last significant single or multi-channel negotiation, kindly evaluate if your negotiation required on a scale from 1 (not a lot of thought) to 5 (a lot of thought).
(2) While thinking of your last significant single or multi-channel negotiation, kindly evaluate if your negotiation required on a scale from 1 (not a lot of focus) to 5 (a lot of focus).
(3) While thinking of your last significant single or multi-channel negotiation, kindly evaluate if your negotiation required on a scale from 1 (not a lot of concentration) to 5 (a lot of concentration).

Source: Adapted from Wilcox et al., 2011

Table 3. Measurement for the Moderating Effect Performance Risk

Performance Risk
(1) There is a high chance that there could be something wrong with a multi-channel negotiation or that the multi-channel negotiation results are not delivered as promised.
(2) There is a high chance of loss in the negotiation because multiple channels are not used well.
(3) A negotiation with multiple channels is extremely risky in terms of how it performs.

Source: Adapted from Keh & Pang, 2010; Stone & Grønhaug, 1993; Dorothea Brack & Benkenstein, 2014

Annex 5. Measurement Scale for Dependent Variables Negotiation Effectiveness and Negotiation Efficiency

Table 1. Measurement for Negotiation Effectiveness

Negotiation Effectiveness
(1) During the negotiation we were able to achieve an optimal outcome regarding the negotiation objects.
(2) The goals defined for the negotiation process were fully achieved.
(3) Examining the current negotiation, the attained negotiation outcome is optimal.
(4) The achieved goal has a positive impact on future relationships with our negotiation partners.
(5) In the current negotiation situation, a better outcome would realistically not be possible.

Source: Barisch, 2011

Table 2. Measurement for Negotiation Efficiency

Negotiation Efficiency
(1) During the negotiation, we only focused on relevant aspects.
(2) All steps of the negotiation were necessary to achieve the target.
(3) Overall, the negotiation was very goal oriented and structured.
(4) A negotiation goal was achieved fast.
(5) The negotiation goal was achieved by involving the least cost / expenses.

Source: Voeth, Herbst & Gawantka, 2006

Annex 6. Demographics

Figure 1. Industry Sectors

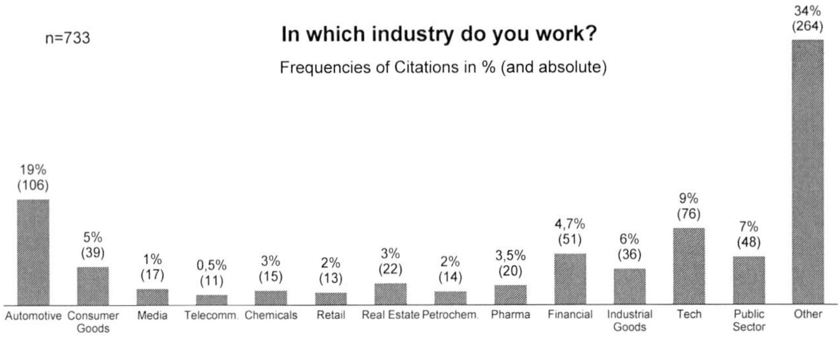

* "Other" Industry: Construction, defense, training & coaching, information technology, legal, consulting

Figure 2. Organizational Size

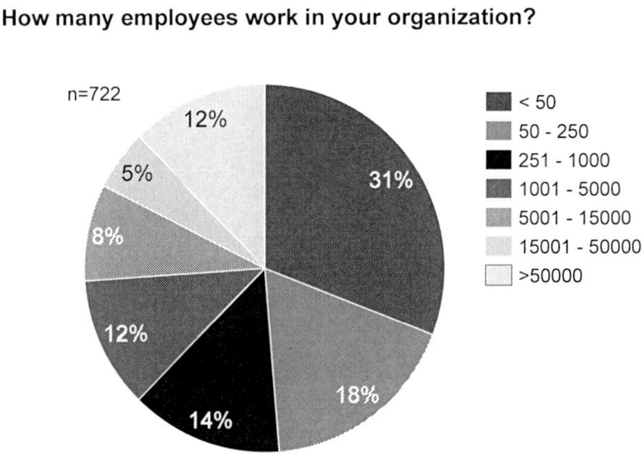

Figure 3. Distribution of Positions Among Survey Participants

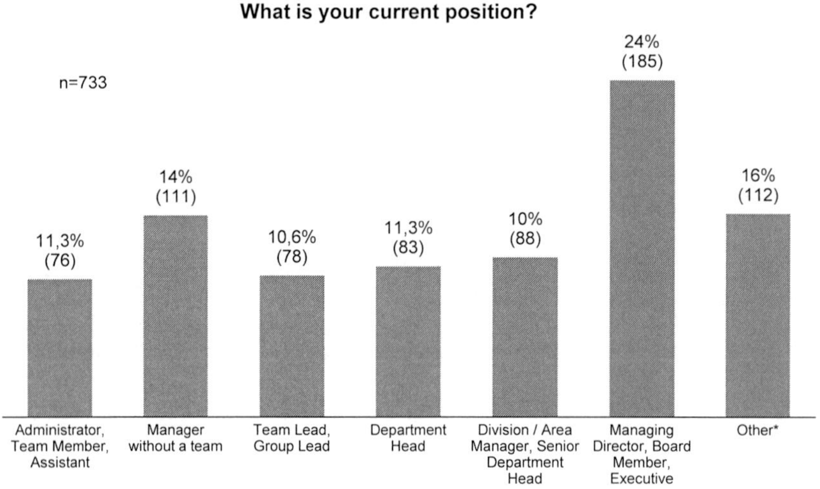

* "Other" Position: Consulting, legal, academia, owner, self-employed, head of union

Annex 7. Qualitative Content Analysis

Table 1. Process Outcome Group Category Negotiation Context

Group	Category	Keywords	#	Free Text
Process Outcome	Negotiation Context	Negotiation Context	495	*Based on high vs. low stakes; Based on how well I know my counterpart; Based on timeline/sense of urgency for me or my counterpart; Based on what each side must gain vs. lose.*
			512	*Depending on the client's preference. Negotiation stage. External restrictions (covid, war).*
			526	*Depending on the context and the sensitivity of the negotiation*
			947	*If possible, I do choose best option, considering the subject, language, cultural and personal factors.*
			984	*The participant numbers, the context. the style, the location of the issue and the negotiators. There are many factors to be considered.*
			1251	*Depending on the context/content of the negotiation - email (or something written) provides a clear document particularly when it is a financial negotiation. If the negotiation is more emotional - in person is my preference.*
			1286	*It depends on the person, their importance, topic, and purpose.*
		Negotiation Partner	64	*It depends on the negotiating partner*
			105	*Depends on the type of the transaction (how serious is it) and depends on the person I'm negotiating with.*
			151	*Depending on the number of counterparts*
		Negotiation Object	123	*The more important the negotiation the more effort I try to put in to have the conversation face-to-face and at least in a video call if f2f is not possible*
			237	*Face to Face to show suppliers the importance of the topic for our organization (especially impactful due to the reduced number of F2F interactions as of COVID-19)*
			374	*I do routine negotiations and because of covid 19, I would prefer using online, like video conference, or phone and WhatsApp. However, if the negotiation is quite important and risky, i choose face to face meeting also attended by various stakeholders.*
			486	*Depends on the volume*
			1336	*Depending on the issues.*
		Complexity	187	*Depending on the complexity of the scope and time constraints*
			640	*In a complex negotiation, email allows accuracy in offers and counteroffers*
			1367	*Based on complexity of information. I include visual for nonverbal clues of understanding.*

Table 2. Process Outcome Group Category Negotiation Management for selection of negotiation channel on purpose 1/2

Group	Category	Keywords	#	Free Text
Process Outcome	Negotiation Management	Strategy & Tactics	101	*Each channel has its own advantages*
			172	*A negotiation plan can be written down and followed more easily via phone*
			205	*Helps me understand the environment & strategize accordingly*
			208	*To influence through different channels*
			250	*Depending on the purpose and strategy as well on the result I want to achieve*
			252	*as Part of tactics*
			398	*This is an important step in planning the negotiation strategy*
			327	*Very few negotiations work well at arm's length. Tidying up details, exchanging (or presenting) versions of a text is usually performed by e-mail. Setting up, and more importantly changing, a schedule belongs on phone.*
			409	*To send a message in between the lines; to show who is in charge*
			435	*All mentioned channels are used most of the time, only the sequence of usage change according to situation*
			445	*Mail as a first contact, phone, video, and WhatsApp to make the follow up. Mail to close and document the agreement.*
			566	*Better power position*

Table 3. Process Outcome Group Category Negotiation Management for selection of negotiation channel on purpose 2/2

Group	Category	Keywords	#	Free Text
Process Outcome	Negotiation Management	Control	57	*Control the environment*
			195	*I can bullet point my negotiation points in a manner which the other party understands, in return they will reply to me back for the points I raised. Ease the communication by being clear and organized.*
			1214	*I select channels control information bandwidth, timing, and emotional impact*
			598	*I like using email to ensure I have all relevant information about the negotiation in front of me. I'm able to clearly outline my questions and review answers in a streamlined fashion.*
		Time Management	69	*For some negotiations I use e-mails because you have more time for work for my BATNA*
			268	*time-management and efficiency*
			338	*To save time, for the first negotiations I always use Conference Calls*
		Convenience	257	*because it is easy to organize, discuss, explain, and negotiate.*
			293	*Specific to the relationship I have built with the other party, I choose what I feel most comfortable with*
			519	*Convenience's sake*
			628	*I'm feeling more comfortable negotiating via email*
			790	*Key point here is that all parties in the negotiation should be comfortable about the channel to be used.*
		Documentation	141	*Email due to advantages of written communication such as rational, unemotional, documented results and offers*
			67	*To have things documented in writing*
			125	*Face-to-face to show commitment and seriousness, written form to summarize and repeat and make sure that opponent has all information clearly spelled out to decide*
			126	*For Accountability purposes and documentation*
			227	*E-Mail for documentation purpose. Video-Calls are easier and convenient due to time saving and travel effort reasons.*
			258	*If possible, I choose e-mail as preferred channel because I have answers, statements etc. in writing. This could be useful as proof if anything goes wrong.*
			371	*Quick result, and recorded*
			457	*No written word should be documented, and the channel must be interactive therefore all the time phone, Face2Face or web conferences, when possible, from both sides*

Table 4. Process Outcome Group Category Behavior for selection of negotiation channel on purpose

Group	Category	Keywords	#	Free Text
Process Outcome	Behavior	Reaction	49	*"via face to face you see the reactions of the sales personal (face, body, movements, talking etc.)"*
		Body Language	56	*"I prefer Face-to-face or Video conferences as I can bring in my personality and can read the counterpart's body language".*
		Emotion	85	*"I favor face to face where possible, to better evaluate the emotional state and responses of my counterparty"*
		Emotion	130	*"Face to face, where you can see directly the reaction/emotion of your negotiation partner"*
		Feeling	262	*"F2F is still the best and most effective way to negotiate since you have a much better feeling for the whole situation. At the end it is the best way to communicate if your negotiation partner is in front of you in reality."*
		Gestures	527	*"I often choose video calls because you can align quickly and can see facial expressions and gestures and it is time saving (no traveling)"*
		Nonverbal cues	587	*"At different stages of the negotiations, it is very important to understand non-verbal communication. This gets lost in channels that are not face-to-face. On the other hand, momenta of legal clarification are great in writing to avoid misunderstands"*
		Nonverbal cues		*"face to face is the only way to collect all non-verbal signals. allows you to add information or go to places at any time to take in visual information"*
		Feeling	662	*"Because you can feel the other person"*
		Reaction, Body Language	825	*"Face to face is more intimate, see reaction and body language, from there you can determine a number of things"*
		Facial expressions, gestures	917	*"I prefer personal contact (face-to-face) because you can build up a different level of relationship and, in particular, also experience facial expressions and gestures in the course of the negotiation and classify them accordingly."*
		Facial expressions, gestures	920	*"Face to face Negotiations are always a best channel as one can read the face expressions and gestures which normally can be overlooked in other channels"*
		Reactions	935	*"I prefer face to face as it allows me to observe and listen more closely to monitor authenticity in communication."*
		Empathize	943	*"It helps to empathize with the other party and feel their mood, tone and stress so that I could make him/her feel at ease"*
		Emotions	956	*"Because it's better to communicate face to face and address any emotions attached"*
		Empathize	1001	*"As an affiliate oriented individual, I am considered an empath. And by analyzing others behavior, I can decide how to reply and answer questions."*
		Body language	1062	*"It mainly depends on English language level of people I negotiate with, however I also like to see a body language during negotiations."*
		Body language	1099	*"face to face enables the opportunity to study communication modes like body language and tone and hence enables confidence evaluation as well."*

Annex 8. Impact of Gender on Channel Choice

Table 1.1. Impact of Gender on the Channel Choice Phone Call

			Female	Male	Total
Phone Call	Not quoted	Quantity	83	191	274
		Expected Quantity	69.2	204.8	274.0
		Standardized Residual	1.7	-1.0	
	Quoted	Quantity	96	339	435
		Expected Quantity	109.8	325.2	435.0
		Standardized Residual	-1.3	.8	
Total		Quantity	179	530	709
		Expected Quantity	179.0	530.0	709.0

Table 1.2. Chi-Square Tests Gender on Selection of Phone Call

	Value	df	Asymptotic Significance (two-sided)	Exact Sig. (two-sided)	Exact Sig. (one-sided)
Pearson-Chi-Square	6.023[a]	1	.014		
Continuation Correction[b]	5.595	1	.018		
Likelihood-Ratio	5.947	1	.015		
Exact Test according to Fisher				.016	.009
Correlation linear-with-linear	6.015	1	.014		
Quantity of valid cases	709				

a. 0 cells (0.0%) display an expected frequency smaller 5. The minimal expected frequency is 69.18.

b. Is only calculated for a 2x2 table

Table 1.3. Symmetric Measures for Gender and Phone Call

		Value	Approximation Significance
Nominal- regarding Nominal Degree	Phi	.092	.014
	Cramer-V	.092	.014
Quantity of valid cases		709	

Table 2.1. Impact of Gender on the Channel Choice Email

			Female	Male	Total
Email	Not quoted	Quantity	56	173	229
		Expected Quantity	57.8	171.2	229.0
		Standardized Residual	-.2	.1	
	Quoted	Quantity	123	357	480
		Expected Quantity	121.2	358.8	480.0
		Standardized Residual	.2	-.1	
Total		Quantity	179	530	709
		Expected Quantity	179.0	530.0	709.0

Table 2.2. Chi-Square Tests Gender on Selection of Email

	Value	df	Asymptotic Significance (two-sided)	Exact Sig. (two-sided)	Exact Sig. (one-sided)
Pearson-Chi-Square	.113[a]	1	.737		
Continuation Correction[b]	.059	1	.808		
Likelihood-Ratio	.113	1	.737		
Exact Test according to Fisher				.782	.406
Correlation linear-with-linear	.112	1	.737		
Quantity of valid cases	709				

a. 0 cells (0.0%) display an expected frequency smaller 5. The minimal expected frequency is 57.82.
b. Is only calculated for a 2x2 table

Table 2.3. Symmetric Measures for Impact of Gender on Email

		Value	Approximation Significance
Nominal- regarding Nominal Degree	Phi	-.013	.737
	Cramer-V	.013	.737
Quantity of valid cases		709	

Table 3.1. Impact of Gender on the Channel Choice Face-to-Face

			Female	Male	Total
Face-to-Face	Not quoted	Quantity	81	168	249
		Expected Quantity	62.9	186.1	249.0
		Standardized Residual	2.3	-1.3	
	Quoted	Quantity	98	362	460
		Expected Quantity	116.1	343.9	460.0
		Standardized Residual	-1.7	1.0	
Total		Quantity	179	530	709
		Expected Quantity	179.0	530.0	709.0

Table 3.2. Chi-Square Tests Gender on Selection of Phone Call

	Value	df	Asymptotic Significance (two-sided)	Exact Sig. (two-sided)	Exact Sig. (one-sided)
Pearson-Chi-Square	10.787[a]	1	.001		
Continuation Correction[b]	10.201	1	.001		
Likelihood-Ratio	10.550	1	.001		
Exact Test according to Fisher				.001	<.001
Correlation linear-with-linear	10.772	1	.001		
Quantity of valid cases	709				

a. 0 cells (0.0%) display an expected frequency smaller 5. The minimal expected frequency is 62.86.

b. Is only calculated for a 2x2 table

Table 3.3. Symmetric Measures

		Value	Approximation Significance
Nominal- regarding Nominal Degree	Phi	.123	.001
	Cramer-V	.123	.001
Quantity of valid cases		709	

Table 4.1. Impact of Gender on the Channel Choice Chat

			Female	Male	Total
Chat	Not quoted	Quantity	147	432	579
		Expected Quantity	146.2	432.8	579.0
		Standardized Residual	.1	.0	
	Quoted	Quantity	32	98	130
		Expected Quantity	32.8	97.2	130.0
		Standardized Residual	-.1	.1	
Total		Quantity	179	530	709
		Expected Quantity	179.0	530.0	709.0

Table 4.2. Chi-Square Tests Gender on Selection of Negotiation Channel Chat

	Value	df	Asymptotic Significance (two-sided)	Exact Sig. (two-sided)	Exact Sig. (one-sided)
Pearson-Chi-Square	.034[a]	1	.854		
Continuation Correction[b]	.005	1	.943		
Likelihood-Ratio	.034	1	.854		
Exact Test according to Fisher				.911	.476
Correlation linear-with-linear	.034	1	.855		
Quantity of valid cases	709				

a. 0 cells (0.0%) display an expected frequency smaller 5. The minimal expected frequency is 32.82.
b. Is only calculated for a 2x2 table

Table 4.3. Symmetric Measures of Gender on Channel Selection Chat

		Value	Approximation Significance
Nominal- regarding Nominal Degree	Phi	.007	.854
	Cramer-V	.007	.854
Quantity of valid cases		709	

Table 5.1. Impact of Gender on the Channel Choice Messaging

			Female	Male	Total
Messaging	Not quoted	Quantity	146	411	557
		Expected Quantity	140.6	416.4	557.0
		Standardized Residual	.5	-.3	
	Quoted	Quantity	33	119	152
		Expected Quantity	38.4	113.6	152.0
		Standardized Residual	-.9	.5	
Total		Quantity	179	530	709
		Expected Quantity	179.0	530.0	709.0

Table 5.2. Chi-Square Tests Gender on Selection of Messaging

	Value	df	Asymptotic Significance (two-sided)	Exact Sig. (two-sided)	Exact Sig. (one-sided)
Pearson-Chi-Square	1.282[a]	1	.258		
Continuation Correction[b]	1.055	1	.304		
Likelihood-Ratio	1.314	1	.252		
Exact Test according to Fisher				.293	.152
Correlation linear-with-linear	1.280	1	.258		
Quantity of valid cases	709				

a. 0 cells (0.0%) display an expected frequency smaller 5. The minimal expected frequency is 38.38.

b. Is only calculated for a 2x2 table

Table 5.3. Symmetric Measures for Impact of Gender on Messaging

		Value	Approximation Significance
Nominal- regarding	Phi	.043	.258
Nominal Degree	Cramer-V	.043	.258
Quantity of valid cases		709	

Table 6.1. Impact of Gender on the Channel Choice Audio Conferencing

			Female	Male	Total
Audio	Not quoted	Quantity	157	445	602
		Expected Quantity	152.0	450.0	602.0
		Standardized Residual	.4	-.2	
	Quoted	Quantity	22	85	107
		Expected Quantity	27.0	80.0	107.0
		Standardized Residual	-1.0	.6	
Total		Quantity	179	530	709
		Expected Quantity	179.0	530.0	709.0

Table 6.2. Chi-Square Tests Gender on Audio Conferencing

	Value	df	Asymptotic Significance (two-sided)	Exact Sig. (two-sided)	Exact Sig. (one-sided)
Pearson-Chi-Square	1.466[a]	1	.226		
Continuation Correction[b]	1.188	1	.276		
Likelihood-Ratio	1.522	1	.217		
Exact Test according to Fisher				.277	.137
Correlation linear-with-linear	1.464	1	.226		
Quantity of valid cases	709				

a. 0 cells (0.0%) display an expected frequency smaller 5. The minimal expected frequency is 27.01.
b. Is only calculated for a 2x2 table

Table 6.3. Symmetric Measures for Impact of Gender on Audio

		Value	Approximation Significance
Nominal- regarding Nominal Degree	Phi	.045	.226
	Cramer-V	.045	.226
Quantity of valid cases		709	

Table 7.1. Impact of Gender on the Channel Choice Video Conferencing

			Female	Male	Total
Video	Not quoted	Quantity	75	188	263
		Expected Quantity	66.4	196.6	263.0
		Standardized Residual	1.1	-.6	
	Quoted	Quantity	104	342	446
		Expected Quantity	112.6	333.4	446.0
		Standardized Residual	-.8	.5	
Total		Quantity	179	530	709
		Expected Quantity	179.0	530.0	709.0

Table 7.2. Chi-Square Tests Gender on Selection of Video Conferencing

	Value	df	Asymptotic Significance (two-sided)	Exact Sig. (two-sided)	Exact Sig. (one-sided)
Pearson-Chi-Square	2.369[a]	1	.124		
Continuation Correction[b]	2.102	1	.147		
Likelihood-Ratio	2.345	1	.126		
Exact Test according to Fisher				.129	.074
Correlation linear-with-linear	2.366	1	.124		
Quantity of valid cases	709				

a. 0 cells (0.0%) display an expected frequency smaller 5. The minimal expected frequency is 66.40.

b. Is only calculated for a 2x2 table

Table 7.3. Symmetric Measures for Impact of Gender on Video

		Value	Approximation Significance
Nominal- regarding Nominal Degree	Phi	.058	.124
	Cramer-V	.058	.124
Quantity of valid cases		709	

Table 8.1. Impact of Gender on the Channel Choice Other

			Female	Male	Total
Phone Call	Not quoted	Quantity	174	510	684
		Expected Quantity	172.7	511.3	684.0
		Standardized Residual	.1	-.1	
	Quoted	Quantity	5	20	25
		Expected Quantity	6.3	18.7	25.0
		Standardized Residual	-.5	.3	
Total		Quantity	179	530	709
		Expected Quantity	179.0	530.0	709.0

Table 8.2. Chi-Square Tests Gender on Selection of Channel Other

	Value	df	Asymptotic Significance (two-sided)	Exact Sig. (two-sided)	Exact Sig. (one-sided)
Pearson-Chi-Square	.378[a]	1	.539		
Continuation Correction[b]	.145	1	.704		
Likelihood-Ratio	.397	1	.529		
Exact Test according to Fisher				.645	.364
Correlation linear-with-linear	.377	1	.539		
Quantity of valid cases	709				

a. 0 cells (0.0%) display an expected frequency smaller 5. The minimal expected frequency is 6.31.
b. Is only calculated for a 2x2 table

Table 8.3. Symmetric Measures for Impact of Gender on the Channel Other

		Value	Approximation Significance
Nominal- regarding Nominal Degree	Phi	.023	.539
	Cramer-V	.023	.539
Quantity of valid cases		709	

Annex 9. Impact of Gender on Channel Quantity

Table 1.1. Impact of Gender on the Quantity of Channels Used

	Gender	N	Mean	Standard Deviation	Standard Error of the Mean Value
Channel Quantity	Female	179	2.8659	1.30853	.09780
	Male	530	3.2491	1.36464	.05928

Table 1.2. Assessment if Gender Impacts the Quantity of Channels through the Test for Independent Samples

		Levene-Test for Variance Equality		T-Test for Average Mean Equality						
		F	Sig.	T	df	Sig. (2-sid.)	Mid Diff.	Diff. Stand. Error	Lower Value	Upper Value
Channel Quantity	Variance is equal	1.460	.227	-3.281	707	.001	-.38313	.11677	-.61239	-.15388
	Variance is not equal			-3.350	318.330	<.001	-.38313	.11436	-.60814	-.15813

Table 1.3. Effect Size of Independent Samples for the Impact of Gender on the Quantity of Channels Used in a Negotiation

		Standardizer[a]	Point Estimation	95% Confidence Interval	
				Lower Value	Upper Value
Choice	Cohen's d	1.35073	-.284	-.454	-.113
	Hedges' Correction	1.35217	-.283	-.453	-.113
	Glass' Delta	1.36464	-.281	-.451	-.110

a. Common denominator used during the estimation of the effect size. Cohen's d uses the summarized standard deviation. Hedges' Correction uses the summarized standard deviation and a correction factor. Glass' Delta uses the standard deviation of a sample from a control group.

Annex 10. Impact of Gender on the Channel Preference

Table 1.1. Impact of Gender on the Channel Preference

			Female	Male	Total
Channel Preference	Phone Call	Quantity	17	40	57
		Expected Quantity	14.4	42.6	57.0
		Standardized Residual	.7	-.4	
	Face-to-Face	Quantity	88	345	433
		Expected Quantity	109.3	323.7	433.0
		Standardized Residual	-2.0	1.2	
	Email	Quantity	23	44	67
		Expected Quantity	16.9	50.1	67.0
		Standardized Residual	1.5	-.9	
	Chat	Quantity	4	3	7
		Expected Quantity	1.8	5.2	7.0
		Standardized Residual	1.7	-1.0	
	Messaging	Quantity	5	12	17
		Expected Quantity	4.3	12.7	17.0
		Standardized Residual	.3	-.2	
	Audio	Quantity	4	9	13
		Expected Quantity	3.3	9.7	13.0
		Standardized Residual	.4	-.2	
	Video	Quantity	35	71	106
		Expected Quantity	26.8	79.2	106.0
		Standardized Residual	1.6	-.9	
	Other	Quantity	3	5	8
		Expected Quantity	2.0	6.0	8.0
		Standardized Residual	.7	-.4	
Total		Quantity	179	530	709
		Expected Quantity	179.0	530.0	709.0

Table 1.2. Impact of Gender on the Preference for a Negotiation Channel

				Monte-Carlo Sign. (two-sided)			Monte-Carlo Sign. (one-sided)		
					99% Confidence Interval			99% Confidence Interval	
	Value	df	Asymp. Signifi. (two-sided)	Sign.	Lower Limit	Upper Limit	Sign.	Lower Limit	Upper Limit
Pearson-Chi-Square	17.629[a]	8	.024	.023[b]	.019	.027			
Likelihood Ratio	17.091	8	.029	.039[b]	.034	.043			
Exact Test Fisher-Freeman-Halton	18.152			.014	.011	.016			
Correlation linear-with-linear	7.026[c]	1	.008	.009[b]	.007	.012	.005[b]	.003	.007
Quantity of valid cases	709								

a. 6 cells (33.3%) display an expected frequency smaller 5. The minimal expected frequency is .25.
b. Based on 10000 sampling tables with a starting value of 1507486128.
c. The standardized statistic is -2.651.

Table 1.3. Symmetric Measures for Gender on the Channel Preference

				Monte-Carlo Significance		
					99% Confidence Interval	
		Value	Approximation Significance	Significance	Lower Limit	Upper Limit
Nominal-regarding Nominal Degree	Phi	.158	.024	.023[a]	.019	.027
	Cramer-V	.158	.024	.023[a]	.019	.027
Quantity of valid cases		709				

a. Based on 10.000 sampling table with a starting value of 1507486128.

Annex 11. Impact of Age on Channel Type

Table 1.1. Group Statistics for Age and Phone Call Use

Group Statistics					
	Phone Call	N	Mean	Std. Deviation	Std. Error Mean
Negotiation Experience	Not quoted	285	27.18	47.719	2.827
	Quoted	448	33.33	39.380	1.861

Table 1.2. Assessment if Experience Impacts Phone Call through the Test for Independent Samples

								Levene-Test Variance Equality		T-Test Average Mean Equality
		F	Sig.	T	df	Sig. (2-sided)	Mid. Diff.	Diff. Stand Error	Lower Value	Upper Value
Channel Quantity	Variance is equal	11.889	0.001	-1.895	731	0.058	-6.148	3.244	-12.516	0.221
	Variance is not equal			-1.817	521.231	0.070	-6.148	3.384	-12.796	0.500

Table 1.3. Effect Size of Independent Samples for the Impact of Negotiation Experience on the Channel Type Phone Call

Independent Samples Effect Sizes

		Standardizer[a]	Point Estimate	95% Confidence Interval	
				Lower	Upper
Negotiation Experience	Cohen's d	42.814	-0.144	-0.292	0.005
	Hedges' Correction	42.858	-0.143	-0.292	0.005
	Glass' Delta	39.380	-0.156	-0.305	-0.007

a. Common denominator used during the estimation of the effect size. Cohen's d uses the summarized standard deviation. Hedges' Correction uses the summarized standard deviation and a correction factor. Glass' Delta uses the standard deviation of a sample from a control group.

Table 2.1. Group Statistics for Age and Face-to-Face Use

Group Statistics

	Phone Call	N	Mean	Std. Deviation	Std. Error Mean
Negotiation Experience	Not quoted	258	28.27	45.756	2.849
	quoted	475	32.39	41.226	1.892

Table 2.2. Assessment if Experience Impacts Face-to-Face Use through the Test for Independent Samples

		Levene-Test Variance Equality		T-Test Average Mean Equality						
		F	Sig.	T	df	Sig. (2-sided)	Mid. Diff.	Diff. Stand. Error	Lower Value	Upper Value
Channel Quantity	Variance is equal	3.493	0.062	-1.244	731	0.214	-4.124	3.316	-10.634	2.385
	Variance is not equal			-1.206	482.716	0.228	-4.124	3.419	-10.843	2.595

Table 2.3. Effect Size of Independent Samples for the Impact of Negotiation Experience on the Channel Type Face-to-Face

Independent Samples Effect Sizes					
		Standardizer[a]	Point Estimate	95% Confidence Interval Lower	Upper
Negotiation Experience	Cohen's d	42.873	-0.096	-0.248	0.055
	Hedges' Correction	42.917	-0.096	-0.248	0.055
	Glass' Delta	41.226	-0.100	-0.252	0.052

a. Common denominator used during the estimation of the effect size. Cohen's d uses the summarized standard deviation. Hedges' Correction uses the summarized standard deviation and a correction factor. Glass' Delta uses the standard deviation of a sample from a control group.

Table 3.1. Group Statistics for Age and Email Use

Group Statistics					
	Phone Call	N	Mean	Std. Deviation	Std. Error Mean
Negotiation Experience	Not quoted	236	31.54	41.219	2.683
	Quoted	497	30.66	43.698	1.960

Table 3.2. Assessment if Experience Impacts Email Use in a Negotiation through the Test for Independent Samples

		Levene-Test for Variance Equality		T-Test for Average Mean Equality						
		F	Sig.	T	df	Sig. (2-sided)	Mid. Diff.	Diff. Stand. Error	Lower Value	Upper Value
Channel Quantity	Variance is equal	0.699	0.403	0.260	731	0.795	0.882	3.393	-5.778	7.543
	Variance is not equal			0.265	487.033	0.791	0.882	3.323	-5.647	7.411

Table 3.3. Effect Size of Independent Samples for the Impact of Negotiation Experience on the Channel Type Email

Independent Samples Effect Sizes

		Standardizer[a]	Point Estimate	95% Confidence Interval Lower	Upper
Negotiation Experience	Cohen's d	42.917	0.021	-0.134	0.175
	Hedges' Correction	42.961	0.021	-0.134	0.175
	Glass' Delta	43.698	0.020	-0.135	0.175

a. Common denominator used during the estimation of the effect size. Cohen's d uses the summarized standard deviation. Hedges' Correction uses the summarized standard deviation and a correction factor. Glass' Delta uses the standard deviation of a sample from a control group.

Table 4.1. Group Statistics for Age and Chat Use

Group Statistics

	Phone Call	N	Mean	Std. Deviation	Std. Error Mean
Negotiation Experience	Not quoted	600	30.42	43.232	1.765
	Quoted	133	33.28	41.388	3.589

Table 4.2. Assessment if Experience Impacts Chat Use in a Negotiation through the Test for Independent Samples

		Levene-Test Variance Equality		T-Test Average Mean Equality						
		F	Sig.	T	df	Sig. (2-sided)	Mid. Diff.	Diff. Stand. Error	Lower Value	Upper Value
Channel Quantity	Variance is equal	0.366	0.545	-0.695	731	0.487	-2.857	4.112	-10.929	5.216
	Variance is not equal			-0.714	200.981	0.476	-2.857	3.999	-10.742	5.029

Table 4.3. Effect Size of Independent Samples for the Impact of Negotiation Experience on the Channel Type Chat

Independent Samples Effect Sizes					
		Standardizer[a]	Point Estimate	95% Confidence Interval Lower	Upper
Negotiation Experience	Cohen's d	42.904	-0.067	-0.254	0.121
	Hedges' Correction	42.949	-0.067	-0.254	0.121
	Glass' Delta	41.388	-0.069	-0.257	0.119

a. Common denominator used during the estimation of the effect size. Cohen's d uses the summarized standard deviation. Hedges' Correction uses the summarized standard deviation and a correction factor. Glass' Delta uses the standard deviation of a sample from a control group.

Table 5.1. Group Statistics for Age and Messaging Use

Group Statistics					
	Phone Call	N	Mean	Std. Deviation	Std. Error Mean
Negotiation Experience	Not quoted	576	30.57	44.281	1.845
	Quoted	157	32.30	37.442	2.988

Table 5.2. Assessment if Experience Impacts Messaging through the Test for Independent Samples

		Levene-Test for Variance Equality		T-Test for Average Mean Equality						
		F	Sig.	T	df	Sig. (2-sided)	Mid. Diff.	Diff. Stand. Error	Lower Value	Upper Value
Channel Quantity	Variance is equal	4.408	0.036	-0.448	731	0.654	-1.730	3.863	-9.315	5.855
	Variance is not equal			-0.493	286.327	0.623	-1.730	3.512	-8.642	5.182

Table 5.3. Effect Size of Independent Samples for the Impact of Negotiation Experience on the Channel Type Messaging

Independent Samples Effect Sizes					
		Standardizer[a]	Point Estimate	95% Confidence Interval Lower	Upper
Negotiation Experience	Cohen's d	42.913	-0.040	-0.217	0.136
	Hedges' Correction	42.957	-0.040	-0.217	0.136
	Glass' Delta	37.442	-0.046	-0.223	0.130

a. Common denominator used during the estimation of the effect size. Cohen's d uses the summarized standard deviation. Hedges' Correction uses the summarized standard deviation and a correction factor. Glass' Delta uses the standard deviation of a sample from a control group.

Table 6.1. Group Statistics for Age and Audio-Conferencing Use

Group Statistics					
	Phone Call	N	Mean	Std. Deviation	Std. Error Mean
Negotiation Experience	Not quoted	622	30.62	43.323	1.737
	Quoted	111	32.73	40.514	3.845

Table 6.2. Assessment if Experience Impacts Audio Conferencing through the Test for Independent Samples

		Levene-Test for Variance Equality		T-Test for Average Mean Equality						
		F	Sig.	T	df	Sig. (2-sided)	Mid. Diff.	Diff. Stand. Error	Lower Value	Upper Value
Channel Quantity	Variance is equal	0.687	0.408	-0.477	731	0.633	-2.109	4.422	-10.790	6.571
	Variance is not equal			-0.500	158.306	0.618	-2.109	4.220	-10.443	6.225

Table 6.3. Effect Size of Independent Samples for the Impact of Negotiation Experience on the Channel Type Audio Conferencing

Independent Samples Effect Sizes

		Standardizer[a]	Point Estimate	95% Confidence Interval	
				Lower	Upper
Negotiation Experience	Cohen's d	42.912	-0.049	-0.251	0.153
	Hedges' Correction	42.956	-0.049	-0.251	0.153
	Glass' Delta	40.514	-0.052	-0.254	0.150

a. Common denominator used during the estimation of the effect size. Cohen's d uses the summarized standard deviation. Hedges' Correction uses the summarized standard deviation and a correction factor. Glass' Delta uses the standard deviation of a sample from a control group.

Table 7.1. Group Statistics for Age and Video Conferencing Use

Group Statistics

	Phone Call	N	Mean	Std. Deviation	Std. Error Mean
Negotiation Experience	Not quoted	275	25.58	48.179	2.905
	Quoted	458	34.16	39.078	1.826

Table 7.2. Assessment if Experience Impacts Video Conferencing through the Test for Independent Samples

		Levene-Test for Variance Equality		T-Test for Average Mean Equality						
		F	Sig.	T	df	Sig. (2-sided)	Mid. Diff.	Diff. Stand. Error	Lower Value	Upper Value
Channel Quantity	Variance is equal	14.340	0.000	-2.631	731	0.009	-8.575	3.259	-14.973	-2.178
	Variance is not equal			-2.499	487.604	0.013	-8.575	3.431	-15.318	-1.833

Table 7.3. Effect Size of Independent Samples for the Impact of Negotiation Experience on the Channel Type Video Conferencing

Independent Samples Effect Sizes					
		Standardizer[a]	Point Estimate	95% Confidence Interval Lower	Upper
Negotiation Experience	Cohen's d	42.717	-0.201	-0.351	-0.051
	Hedges' Correction	42.761	-0.201	-0.350	-0.051
	Glass' Delta	39.078	-0.219	-0.370	-0.069

a. Common denominator used during the estimation of the effect size. Cohen's d uses the summarized standard deviation. Hedges' Correction uses the summarized standard deviation and a correction factor. Glass' Delta uses the standard deviation of a sample from a control group.

Table 8.1. Group Statistics for Age and Alternative Channel Use

Group Statistics					
	Phone Call	N	Mean	Std. Deviation	Std. Error Mean
Negotiation Experience	Not quoted	707	30.81	43.002	1.617
	Quoted	26	34.42	40.326	7.909

Table 8.2. Assessment if Experience Impacts Alternative Channel through the Test for Independent Samples

		Levene-Test for Variance Equality		T-Test for Average Mean Equality						
		F	Sig.	T	df	Sig. (2-sided)	Mid. Diff.	Diff. Stand. Error	Lower Value	Upper Value
Channel Quantity	Variance is equal	0.234	0.629	-0.421	731	0.674	-3.611	8.569	-20.435	13.212
	Variance is not equal			-0.447	27.133	0.658	-3.611	8.072	-20.170	12.948

Table 8.3. Effect Size of Independent Samples for the Impact of Negotiation Experience on the Channel Type Other

Independent Samples Effect Sizes					
		Standardizer[a]	Point Estimate	95% Confidence Interval Lower	Upper
Negotiation Experience	Cohen's d	42.913	-0.084	-0.476	0.307
	Hedges' Correction	42.957	-0.084	-0.475	0.307
	Glass' Delta	40.326	-0.090	-0.481	0.303

a. Common denominator used during the estimation of the effect size. Cohen's d uses the summarized standard deviation. Hedges' Correction uses the summarized standard deviation and a correction factor. Glass' Delta uses the standard deviation of a sample from a control group.

Annex 12. Correlation between Age and Quantity of Channels

Table 1. Correlation between Age and Quantity of Channels

	Correlations	Channel Quantity
Kindly let us know your age in the box below.	Pearson Correlation	.091*
	Sig. (two-tailed)	.014
	N	733

*The correlation is significant on a level of 0.05 (two-tailed).

Annex 13. Impact of Industry on the Type of Channels Used

Table 1.1. Phone Call * Industry Cross Table

		Phone Call * Industry Cross Table														
		1	2	3	4	5	6	7	8	9	10	11	12	13	14	Total
Quoted	Quantity	43	14	6	3	4	4	7	5	10	17	14	38	27	93	285
	Expected Quantity	41.2	15.2	6.6	4.3	5.8	5.1	8.6	5.4	7.8	19.8	14.0	29.5	18.7	102.6	285.0
	Standardized Residual	0.3	-0.3	-0.2	-0.6	-0.8	-0.5	-0.5	-0.2	0.8	-0.6	0.0	1.6	1.9	-1.0	
Not quoted	Quantity	63	25	11	8	11	9	15	9	10	34	22	38	21	171	448
	Expected Quantity	64.8	23.8	10.4	6.7	9.2	7.9	13.4	8.6	12.2	31.2	22.0	46.5	29.3	161.4	448.0
	Standardized Residual	-0.2	0.2	0.2	0.5	0.6	0.4	0.4	0.2	-0.6	0.5	0.0	-1.2	-1.5	0.8	
	Quantity	106	39	17	11	15	13	22	14	20	51	36	76	48	264	733
	Expected Quantity	106.0	39.0	17.0	11.0	15.0	13.0	22.0	14.0	20.0	51.0	36.0	76.0	48.0	264.0	733.0

Note: Automotive (1), Consumer Goods (2), Media (3), Telecommunications (4), Chemicals (5), Retail (6), Real Estate (7), Petrochemical Sector / Oil Industry (8), Pharmaceuticals (9), Financial Services (10), Industrial Goods (11), High Technology Industry (12), Public Sector (13), Other (14)

Table 1.2. Chi-Square Tests for Industry and Phone Call

	Value	df	Asymptotic Significance (2-sided)
	16.679[a]	14	0.274
	16.847	14	0.264
	0.196	1	0.658
	733		

a. 3 cells (10.0%) display an estimated frequency smaller 5. The minimal expected frequency is .39.

Table 1.3. Symmetric Measures

		Symmetric Measures	
		Value	Approximate Significance
Nominal by nominal	Phi	0.151	0.274
	Cramer-V	0.151	0.274
Valid cases		733	

Table 2.1. Face-to-Face * Industry Cross Table

		Face-to-Face * Industry Cross Table														
		1	2	3	4	5	6	7	8	9	10	11	12	13	14	Total
Quoted	Quantity	35	16	9	4	5	3	5	6	3	19	8	30	15	99	258
	Expected Quantity	37.3	13.7	6.0	3.9	5.3	4.6	7.7	4.9	7.0	18.0	12.7	26.8	16.9	92.9	258.0
	Standardized Residual	-0.4	0.6	1.2	0.1	-0.1	-0.7	-1.0	0.5	-1.5	0.2	-1.3	0.6	-0.5	0.6	
Not quoted	Quantity	71	23	8	7	10	10	17	8	17	32	28	46	33	165	475
	Expected Quantity	68.7	25.3	11.0	7.1	9.7	8.4	14.3	9.1	13.0	33.0	23.3	49.2	31.1	171.1	475.0
	Standardized Residual	0.3	-0.5	-0.9	0.0	0.1	0.5	0.7	-0.4	1.1	-0.2	1.0	-0.5	0.3	-0.5	
	Quantity	35	16	9	4	5	3	5	6	3	19	8	30	15	99	258
	Expected Quantity	37.3	13.7	6.0	3.9	5.3	4.6	7.7	4.9	7.0	18.0	12.7	26.8	16.9	92.9	258.0

Note: Automotive (1), Consumer Goods (2), Media (3), Telecommunications (4), Chemicals (5), Retail (6), Real Estate (7), Petrochemical Sector / Oil Industry (8), Pharmaceuticals (9), Financial Services (10), Industrial Goods (11), High Technology Industry (12), Public Sector (13), Other (14)

Table 2.2. Chi-Quadrat-Tests

	Value	df	Asymptotic Significance (2-sided)
	15.595[a]	14	0.339
	16.598	14	0.278
	0.038	1	0.846
	733		

a. 5 cells (16.7%) display an expected frequency smaller 5. The minimal expected frequency is .35.

Table 2.3. Symmetric Measures

Symmetric Measures			
		Value	Approximate Significance
Nominal by Nominal	Phi	0.146	0.339
	Cramer-V	0.146	0.339
Valid cases		733	

Table 3.1. Email * Industry Cross Table

		Email * Industry Cross Table														
		1	2	3	4	5	6	7	8	9	10	11	12	13	14	Total
Quoted	Quantity	45	9	3	2	2	2	8	0	10	21	10	26	18	79	236
	Expected Quantity	34.1	12.6	5.5	3.5	4.8	4.2	7.1	4.5	6.4	16.4	11.6	24.5	15.5	85.0	236.0
	Standardized Residual	1.9	-1.0	-1.1	-0.8	-1.3	-1.1	0.3	-2.1	1.4	1.1	-0.5	0.3	0.6	-0.7	
Not quoted	Quantity	61	30	14	9	13	11	14	14	10	30	26	50	30	185	497
	Expected Quantity	71.9	26.4	11.5	7.5	10.2	8.8	14.9	9.5	13.6	34.6	24.4	51.5	32.5	179.0	497.0
	Standardized Residual	-1.3	0.7	0.7	0.6	0.9	0.7	-0.2	1.5	-1.0	-0.8	0.3	-0.2	-0.4	0.4	
	Quantity	106	39	17	11	15	13	22	14	20	51	36	76	48	264	733
	Expected Quantity	106.0	39.0	17.0	11.0	15.0	13.0	22.0	14.0	20.0	51.0	36.0	76.0	48.0	264.0	733.0

Note: Automotive (1), Consumer Goods (2), Media (3), Telecommunications (4), Chemicals (5), Retail (6), Real Estate (7), Petrochemical Sector / Oil Industry (8), Pharmaceuticals (9), Financial Services (10), Industrial Goods (11), High Technology Industry (12), Public Sector (13), Other (14)

Table 3.2. Chi-Quadrat-Tests

	Value	df	Asymptotic Significance (2-sided)
	28.782[a]	14	0.011
	33.705	14	0.002
	0.382	1	0.537
	733		

a. 6 cells (20.0%) display an expected frequency smaller 5. The minimal expected frequency is .32.

Table 3.3. Symmetric Measures

Symmetric Measures			
		Value	Approximate Significance
Nominal by Nominal	Phi	0.198	0.011
	Cramer-V	0.198	0.011
Valid cases		733	

Table 4.1. Chat * Industry Cross Table

Chat * Industry Cross Table																
		1	2	3	4	5	6	7	8	9	10	11	12	13	14	Total
Quoted	Quantity	89	31	14	8	14	9	18	12	13	40	28	61	43	219	600
	Expected Quantity	86.8	31.9	13.9	9.0	12.3	10.6	18.0	11.5	16.4	41.7	29.5	62.2	39.3	216.1	600.0
	Standardized Residual	0.2	-0.2	0.0	-0.3	0.5	-0.5	0.0	0.2	-0.8	-0.3	-0.3	-0.2	0.6	0.2	
Not quoted	Quantity	17	8	3	3	1	4	4	2	7	11	8	15	5	45	133
	Expected Quantity	19.2	7.1	3.1	2.0	2.7	2.4	4.0	2.5	3.6	9.3	6.5	13.8	8.7	47.9	133.0
	Standardized Residual	-0.5	0.3	0.0	0.7	-1.0	1.1	0.0	-0.3	1.8	0.6	0.6	0.3	-1.3	-0.4	
	Quantity	106	39	17	11	15	13	22	14	20	51	36	76	48	264	733
	Expected Quantity	106.0	39.0	17.0	11.0	15.0	13.0	22.0	14.0	20.0	51.0	36.0	76.0	48.0	264.0	733.0

Note: Automotive (1), Consumer Goods (2), Media (3), Telecommunications (4), Chemicals (5), Retail (6), Real Estate (7), Petrochemical Sector / Oil Industry (8), Pharmaceuticals (9), Financial Services (10), Industrial Goods (11), High Technology Industry (12), Public Sector (13), Other (14)

Table 4.2. Chi-Square-Tests

	Value	df	Asymptotic Significance (2-sided)
	11.077[a]	14	0.680
	10.981	14	0.688
	0.037	1	0.847
	733		

a. 9 cells (30.0%) display an expected frequency smaller 5. The minimal expected frequency is .18.

Table 4.3. Symmetric Measures

	Symmetric Measures	Value	Approximate Significance
Nominal by Nominal	Phi	0.123	0.680
	Cramer-V	0.123	0.680
Valid cases		733	

Table 5.1. Messaging * Industry Cross Table

		Messaging * Industry Cross Table														
		1	2	3	4	5	6	7	8	9	10	11	12	13	14	Total
Quoted	Quantity	87	27	12	8	12	10	11	11	16	39	30	64	40	209	576
	Expected Quantity	83.3	30.6	13.4	8.6	11.8	10.2	17.3	11.0	15.7	40.1	28.3	59.7	37.7	207.5	576.0
	Standardized Residual	0.4	-0.7	-0.4	-0.2	0.1	-0.1	-1.5	0.0	0.1	-0.2	0.3	0.6	0.4	0.1	
Not quoted	Quantity	19	12	5	3	3	3	11	3	4	12	6	12	8	55	157
	Expected Quantity	22.7	8.4	3.6	2.4	3.2	2.8	4.7	3.0	4.3	10.9	7.7	16.3	10.3	56.5	157.0
	Standardized Residual	-0.8	1.3	0.7	0.4	-0.1	0.1	2.9	0.0	-0.1	0.3	-0.6	-1.1	-0.7	-0.2	
	Quantity	106	39	17	11	15	13	22	14	20	51	36	76	48	264	733
	Expected Quantity	106.0	39.0	17.0	11.0	15.0	13.0	22.0	14.0	20.0	51.0	36.0	76.0	48.0	264.0	733.0

Note: Automotive (1), Consumer Goods (2), Media (3), Telecommunications (4), Chemicals (5), Retail (6), Real Estate (7), Petrochemical Sector / Oil Industry (8), Pharmaceuticals (9), Financial Services (10), Industrial Goods (11), High Technology Industry (12), Public Sector (13), Other (14)

Table 5.2. Chi-Quadrat-Tests

	Value	df	Asymptotic Significance (2-sided)
	20.820[a]	14	0.106
	18.233	14	0.196
	1.129	1	0.288
	733		

a. 9 cells (30.0%) display an expected frequency smaller 5. The minimal expected frequency is .18.

Table 5.3. Symmetric Measures

	Symmetric Measures	Value	Approximate Significance
Nominal by Nominal	Phi	0.169	0.106
	Cramer-V	0.169	0.106
Valid cases		733	

Table 6.1. Audio Conferencing * Industry Cross Table

		Audio Conferencing * Industry Cross Table														
		1	2	3	4	5	6	7	8	9	10	11	12	13	14	Total
Quoted	Quantity	80	33	15	9	13	13	18	10	16	39	31	66	43	235	622
	Expected Quantity	89.9	33.1	14.4	9.3	12.7	11.0	18.7	11.9	17.0	43.3	30.5	64.5	40.7	224.0	622.0
	Standardized Residual	-1.0	0.0	0.2	-0.1	0.1	0.6	-0.2	-0.5	-0.2	-0.7	0.1	0.2	0.4	0.7	
Not quoted	Quantity	26	6	2	2	2	0	4	4	4	12	5	10	5	29	111
	Expected Quantity	16.1	5.9	2.6	1.7	2.3	2.0	3.3	2.1	3.0	7.7	5.5	11.5	7.3	40.0	111.0
	Standardized Residual	2.5	0.0	-0.4	0.3	-0.2	-1.4	0.4	1.3	0.6	1.5	-0.2	-0.4	-0.8	-1.7	
	Quantity	106	39	17	11	15	13	22	14	20	51	36	76	48	264	733
	Expected Quantity	106.0	39.0	17.0	11.0	15.0	13.0	22.0	14.0	20.0	51.0	36.0	76.0	48.0	264.0	733.0

Note: Automotive (1), Consumer Goods (2), Media (3), Telecommunications (4), Chemicals (5), Retail (6), Real Estate (7), Petrochemical Sector / Oil Industry (8), Pharmaceuticals (9), Financial Services (10), Industrial Goods (11), High Technology Industry (12), Public Sector (13), Other (14)

Table 6.2. Chi-Square-Tests

	Value	df	Asymptotic Significance (2-sided)
	19.979[a]	14	0.131
	20.857	14	0.105
	8.217	1	0.004
	733		

a. 9 cells (30.0%) display an expected frequency smaller 5. The minimal expected frequency is .15.

Table 6.3: Symmetric Measures

	Symmetric Measures	Value	Approximate Significance
Nominal by Nominal	Phi	0.165	0.131
	Cramer-V	0.165	0.131
Valid cases		733	

Table 7.1. Video Conferencing * Industry Cross Table

		Video Conferencing * Industry Cross Table														
		1	2	3	4	5	6	7	8	9	10	11	12	13	14	Total
Quoted	Quantity	38	14	4	4	4	3	12	9	11	16	8	27	21	104	275
	Expected Quantity	39.8	14.6	6.4	4.1	5.6	4.9	8.3	5.3	7.5	19.1	13.5	28.5	18.0	99.0	275.0
	Standardized Residual	-0.3	-0.2	-0.9	-0.1	-0.7	-0.9	1.3	1.6	1.3	-0.7	-1.5	-0.3	0.7	0.5	
Not quoted	Quantity	68	25	13	7	11	10	10	5	9	35	28	49	27	160	458
	Expected Quantity	66.2	24.4	10.6	6.9	9.4	8.1	13.7	8.7	12.5	31.9	22.5	47.5	30.0	165.0	458.0
	Standardized Residual	0.2	0.1	0.7	0.0	0.5	0.7	-1.0	-1.3	-1.0	0.6	1.2	0.2	-0.5	-0.4	
	Quantity	106	39	17	11	15	13	22	14	20	51	36	76	48	264	733
	Expected Quantity	106.0	39.0	17.0	11.0	15.0	13.0	22.0	14.0	20.0	51.0	36.0	76.0	48.0	264.0	733.0

Note: Automotive (1), Consumer Goods (2), Media (3), Telecommunications (4), Chemicals (5), Retail (6), Real Estate (7), Petrochemical Sector / Oil Industry (8), Pharmaceuticals (9), Financial Services (10), Industrial Goods (11), High Technology Industry (12), Public Sector (13), Other (14)

Table 7.2. Chi-Quadrat-Tests

	Value	df	Asymptotic Significance (2-sided)
	19.448[a]	14	0.149
	19.927	14	0.132
	0.732	1	0.392
	733		

a. 9 cells (30.0%) display an expected frequency less than 5. The minimal expected frequency is .15.

Table 7.3. Symmetric Measures

	Symmetric Measures		
		Value	Approximate Significance
Nominal by Nominal	Phi	0.163	0.149
	Cramer-V	0.163	0.149
Valid cases		733	

Note: Due to the low selection of “other” channel by survey respondents this category is not analyzed.

Annex 14. Impact of Industry on the Channel Quantity

Table 1.1. Homogeneity Variances Test for Industry and Channel Quantity

		Tests of Homogeneity of Variances			
		Levene Statistic	df1	df2	Sig.
Channel Quantity	Based on Mean	0.592	13	718	0.862
	Based on Median	0.435	13	718	0.957
	Based on Median and with adjusted df	0.435	13	685.689	0.957
	Based on trimmed mean	0.565	13	718	0.882

Table 1.2. ANOVA for Industry and Channel Quantity

		ANOVA				
		Sum of Squares	df	Mean Square	F	Sig.
Channel Quantity	Between Groups	29.773	14	2.127	1.144	0.315
	Within Groups	1334.898	718	1.859		
	Total	1364.671	732			

Annex 15. Impact of Position on Channel Type

Table 1.1. Phone Call * Position Cross Tabulation

Phone Call * Position Cross Table									
		Administrator, Team Member, Assistant	Manager without a team	Team Lead, Group Lead	Department Head	Division Manager, Area Manager, Business Unit Manager, Senior Department Head	Managing Director, Board Member, Executive	Other	Total
Quoted	Quantity	35	50	38	28	33	63	38	285
	Expected Quantity	29.5	43.2	30.3	32.3	34.2	71.9	43.5	285.0
	Standardized Residual	1.0	1.0	1.4	-.8	-.2	-1.1	-.8	
Not quoted	Quantity	41	61	40	55	55	122	74	448
	Expected Quantity	46.5	67.8	47.7	50.7	53.8	113.1	68.5	448.0
	Standardized Residual	-.8	-.8	-1.1	.6	.2	.8	.7	
	Quantity	76	111	78	83	88	185	112	733
	Expected Quantity	76.0	111.0	78.0	83.0	88.0	185.0	112.0	733.0

Table 1.2. Chi-Square Tests for Position and Phone Call

	Value	df	Asymptotic Significance (2-sided)
Pearson Chi-Square	10.561[a]	6	.103
Likelihood Ratio	10.498	6	.105
Linear-by-Linear Association	7.692	1	.006
N of Valid Cases	733		

a. 0 cells (0.0%) display an expected count less than 5. The minimal expected count is 29.55.

Table 1.3. Symmetric Measures for Position and Phone Call

Symmetric Measures			
		Value	Approximate Significance
Nominal by Nominal	Phi	.120	.103
	Cramer's V	.120	.103
N of Valid Cases		733	

Table 2.1. Face-to-Face * Position Cross Tabulation

		Phone Call * Position Cross Table							
		Administrator, Team Member, Assistant	Manager without a team	Team Lead, Group Lead	Department Head	Division Manager, Area Manager, Business Unit Manager, Senior Department Head	Managing Director, Board Member, Executive	Other	Total
Quoted	Quantity	25	48	32	26	28	57	42	258
	Expected Quantity	26.8	39.1	27.5	29.2	31.0	65.1	39.4	258.0
	Standardized Residual	-.3	1.4	.9	-.6	-.5	-1.0	.4	
Not quoted	Quantity	51	63	46	57	60	128	70	475
	Expected Quantity	49.2	71.9	50.5	53.8	57.0	119.9	72.6	475.0
	Standardized Residual	.2	-1.1	-.6	.4	.4	.7	-.3	
	Quantity	76	111	78	83	88	185	112	733
	Expected Quantity	76.0	111.0	78.0	83.0	88.0	185.0	112.0	733.0

Table 2.2. Chi-Square Tests for Position and Face-to-Face

	Value	df	Asymptotic Significance (2-sided)
Pearson Chi-Square	7.296[a]	6	.294
Likelihood Ratio	7.229	6	.300
Linear-by-Linear Association	1.218	1	.270
N of Valid Cases	733		

a. 0 cells (0.0%) display an expected count less than 5. The minimal expected count is 26.75.

Table 2.3. Symmetric Measures for Position and Face-to-Face

Symmetric Measures for Position and Face-to-Face			
		Value	Approximate Significance
Nominal by Nominal	Phi	.100	.294
	Cramer's V	.100	.294
N of Valid Cases		733	

Table 3.1. Email * Position Cross Tabulation

		Email * Position Cross Table							
		Administrator, Team Member, Assistant	Manager without a team	Team Lead, Group Lead	Department Head	Division Manager, Area Manager, Business Unit Manager, Senior Department Head	Managing Director, Board Member, Executive	Other	Total
Quoted	Quantity	24	41	27	21	29	67	27	236
	Expected Quantity	24.5	35.7	25.1	26.7	28.3	59.6	36.1	236.0
	Standardized Residual	-.1	.9	.4	-1.1	.1	1.0	-1.5	
Not quoted	Quantity	52	70	51	62	59	118	85	497
	Expected Quantity	51.5	75.3	52.9	56.3	59.7	125.4	75.9	497.0
	Standardized Residual	.1	-.6	-.3	.8	-.1	-.7	1.0	
	Quantity	76	111	78	83	88	185	112	733
	Expected Quantity	76.0	111.0	78.0	83.0	88.0	185.0	112.0	733.0

Table 3.2. Chi-Quadrat-Tests for Position and Email

	Value	df	Asymptotic Significance (2-sided)
Pearson Chi-Square	7.922[a]	6	.244
Likelihood Ratio	8.124	6	.229
Linear-by-Linear Association	.796	1	.372
N of Valid Cases	733		

a. 0 cells (0.0%) display an expected count less than 5. The minimal expected count is 24.47.

Table 3.3. Symmetric Measures for Position and Email

Symmetric Measures

		Value	Approximate Significance
Nominal by Nominal	Phi	.104	.244
	Cramer's V	.104	.244
N of Valid Cases		733	

Table 4.1. Chat * Position Crosstabulation

Chat * Position Cross Table									
		Administrator, Team Member, Assistant	Manager without a team	Team Lead, Group Lead	Department Head	Division Manager, Area Manager, Business Unit Manager, Senior Department Head	Managing Director, Board Member, Executive	Other	Total
Quoted	Quantity	57	90	67	67	72	151	96	600
	Expected Quantity	62.2	90.9	63.8	67.9	72.0	151.4	91.7	600.0
	Standardized Residual	-.7	-.1	.4	-.1	.0	.0	.5	
Not quoted	Quantity	19	21	11	16	16	34	16	133
	Expected Quantity	13.8	20.1	14.2	15.1	16.0	33.6	20.3	133.0
	Standardized Residual	1.4	.2	-.8	.2	.0	.1	-1.0	
	Quantity	76	111	78	83	88	185	112	733
	Expected Quantity	76.0	111.0	78.0	83.0	88.0	185.0	112.0	733.0

Table 4.2. Chi-Square Tests for Position and Chat

	Value	df	Asymptotic Significance (2-sided)
Pearson Chi-Square	4.509[a]	6	.608
Likelihood Ratio	4.437	6	.618
Linear-by-Linear Association	1.575	1	.210
N of Valid Cases	733		

a. 0 cells (0.0%) display an expected count less than 5. The minimal expected count is 13.79.

Table 4.3. Symmetric Measures for Position and Chat

Symmetric Measures

		Value	Approximate Significance
Nominal by Nominal	Phi	.078	.608
	Cramer's V	.078	.608
N of Valid Cases		733	

Table 5.1. Messaging * Position Cross Tabulation

Messaging * Position Cross Table

		Administrator, Team Member, Assistant	Manager without a team	Team Lead, Group Lead	Department Head	Division Manager, Area Manager, Business Unit Manager, Senior Department Head	Managing Director, Board Member, Executive	Other	Total
Quoted	Quantity	63	88	65	65	73	132	90	576
	Expected Quantity	59.7	87.2	61.3	65.2	69.2	145.4	88.0	576.0
	Standardized Residual	.4	.1	.5	.0	.5	-1.1	.2	
Not quoted	Quantity	13	23	13	18	15	53	22	157
	Expected Quantity	16.3	23.8	16.7	17.8	18.8	39.6	24.0	157.0
	Standardized Residual	-.8	-.2	-.9	.1	-.9	2.1	-.4	
	Quantity	76	111	78	83	88	185	112	733
	Expected Quantity	76.0	111.0	78.0	83.0	88.0	185.0	112.0	733.0

Table 5.2. Chi-Square Tests for Position and Messaging

	Value	df	Asymptotic Significance (2-sided)
Pearson Chi-Square	8.878[a]	6	0.181
Likelihood Ratio	8.652	6	0.194
Linear-by-Linear Association	2.066	1	0.151
N of Valid Cases	733		

a. 0 cells (0.0%) display an expected count less than 5. The minimal expected count is 16.28.

Table 5.3. Symmetric Measures for Position and Messaging

Symmetric Measures

		Value	Approximate Significance
Nominal by Nominal	Phi	.110	.181
	Cramer's V	.110	.181
N of Valid Cases		733	

Table 6.1. Audio Conferencing * Position Cross Table

		Audio Conferencing * Position Cross Table							
		Administrator, Team Member, Assistant	Manager without a team	Team Lead, Group Lead	Department Head	Division Manager, Area Manager, Business Unit Manager, Senior Department Head	Managing Director, Board Member, Executive	Other	Total
Quoted	Quantity	65	92	68	59	80	160	98	622
	Expected Quantity	64.5	94.2	66.2	70.4	74.7	157.0	95.0	622.0
	Standardized Residual	.1	-.2	.2	-1.4	.6	.2	.3	
Not quoted	Quantity	11	19	10	24	8	25	14	111
	Expected Quantity	11.5	16.8	11.8	12.6	13.3	28.0	17.0	111.0
	Standardized Residual	-.1	.5	-.5	3.2	-1.5	-.6	-.7	
	Quantity	76	111	78	83	88	185	112	733
	Expected Quantity	76.0	111.0	78.0	83.0	88.0	185.0	112.0	733.0

Table 6.2. Chi-Square Tests for Position and Audio Conferencing

	Value	df	Asymptotic Significance (2-sided)
Pearson Chi-Square	16.442[a]	6	.012
Likelihood Ratio	14.739	6	.022
Linear-by-Linear Association	1.195	1	.274
N of Valid Cases	733		

a. 0 cells (0.0%) display an expected count less than 5. The minimal expected count is 11.51.

Table 6.3. Symmetric Measures for Position and Audio Conferencing

	Symmetric Measures	Value	Approximate Significance
Nominal by Nominal	Phi	.150	.012
	Cramer's V	.150	.012
N of Valid Cases		733	

Table 7.1. Video Conferencing * Position Cross Table

		Video Conferencing * Position Cross Table							
		Administrator, Team Member, Assistant	Manager without a team	Team Lead, Group Lead	Department Head	Division Manager, Area Manager, Business Unit Manager, Senior Department Head	Managing Director, Board Member, Executive	Other	Total
Quoted	Quantity	29	41	32	28	22	71	52	275
	Expected Quantity	28.5	41.6	29.3	31.1	33.0	69.4	42.0	275.0
	Standardized Residual	0.1	-0.1	0.5	-0.6	-1.9	0.2	1.5	
Not quoted	Quantity	47	70	46	55	66	114	60	458
	Expected Quantity	47.5	69.4	48.7	51.9	55.0	115.6	70.0	458.0
	Standardized Residual	-0.1	0.1	-0.4	0.4	1.5	-0.1	-1.2	
	Quantity	76	111	78	83	88	185	112	733
	Expected Quantity	76.0	111.0	78.0	83.0	88.0	185.0	112.0	733.0

Table 7.2. Chi-Square Tests for Position and Video Conferencing

	Value	df	Asymptotic Significance (2-sided)
Pearson Chi-Square	10.680[a]	6	.099
Likelihood Ratio	10.948	6	.090
Linear-by-Linear Association	.533	1	.465
N of Valid Cases	733		

a. 0 cells (0.0%) display an expected count less than 5. The minimal expected count is 28.51.

Table 7.3. Symmetric Measures for Position and Video Conferencing

Symmetric Measures

		Value	Approximate Significance
Nominal by Nominal	Phi	.121	.099
	Cramer's V	.121	.099
N of Valid Cases		733	

Note: Due to the low selection of "other" channel by survey respondents this category is not analyzed.

Annex 16. Impact of Position on Channel Quantity

Table 1.1. Tests of Homogeneity of Variances

Tests of Homogeneity of Variances

		Levene Statistic	df1	df2	Sig.
Channel Quantity	Based on Mean	1.365	6	726	.226
	Based on Median	1.317	6	726	.247
	Based on Median and with adjusted df	1.317	6	697,271	.247
	Based on trimmed mean	1.488	6	726	.179

Table 1.2. ANOVA for Position and Channel Quantity

ANOVA

		Sum of Squares	df	Mean Square	F	Sig.
Channel Quantity	Between Groups	26.156	6	4.359	2.364	.029
	Within Groups	1338.515	726	1.844		
	Total	1364.671	732			

Table 1.3. ANOVA Effect Sizes for Position and Channel Quantity

ANOVA Effect Sizes[a,b]

			95% Confidence Interval	
Choice Count		Point Estimation	Lower	Higher
	Eta-Square	.019	.000	.035
	Epsilon-Square	.011	-.008	.027
	Omega-Square, solid Effect	.011	-.008	.027
	Omega-Square, Random Effect	.002	-.001	.005

a. Eta-Square and Epsilon-Square are estimated based on this model with solid effects.
b. Negative and lesser skewed estimations are kept and not rounded off to zero.

Table 1.4. Multiple Comparisons for Position and Channel Quantity 1/2

Multiple Comparisons * Dependent Variable * Channel Quantity * Scheffe

(I) Position	(J) Position	Mean Difference (I-J)	Std. Error	Sig.	95% Confidence Interval	
					Lower Bound	Upper Bound
Administrator, Team Member..	Manager without a team	.12328	.20216	.999	-.5962	.8428
	Team Lead, Group Lead	.29757	.21885	.933	-.4813	1.0765
	Department Head	-.41281	.21557	.722	-1.1800	.3544
	Division Manager etc.	-.08792	.21263	1.000	-.8447	.6688
	Managing Director, Board Member, Executive	-.14339	.18500	.996	-.8018	.5150
	Other	.02491	.20179	1.000	-.6933	.7431
Manager without a team	Administrator, Team Member..	-.12328	.20216	.999	-.8428	.5962
	Team Lead, Group Lead	.17429	.20062	.993	-.5397	.8883
	Department Head	-.53609	.19704	.287	-1.2373	.1652
	Division Manager etc.	-.21120	.19381	.977	-.9010	.4786
	Managing Director, Board Member, Executive	-.26667	.16302	.848	-.8469	.3135
	Other	-.09838	.18185	1.000	-.7456	.5488
Team Lead, Group Lead	Administrator, Team Member..	-.29757	.21885	.933	-1.0765	.4813
	Manager without a team	-.17429	.20062	.993	-.8883	.5397
	Department Head	-.71038	.21413	.090	-1.4724	.0517
	Division Manager etc.	-.38549	.21116	.766	-1.1370	.3660
	Managing Director, Board Member, Executive	-.44096	.18331	.448	-1.0934	.2114
	Other	-.27266	.20025	.932	-.9853	.4400

Table 1.5. Multiple Comparisons for Position and Channel Quantity 2/2

Multiple Comparisons * Dependent Variable * Channel Quantity * Scheffe

(I) Position	(J) Position	Mean Difference (I-J)	Std. Error	Sig.	95% Confidence Interval	
					Lower Bound	Upper Bound
Department Head	Administrator, Team Member..	.41281	.21557	.722	-.3544	1.1800
	Manager without a team	.53609	.19704	.287	-.1652	1.2373
	Team Lead, Group Lead	.71038	.21413	.090	-.0517	1.4724
	Division Manager etc.	.32489	.20776	.874	-.4145	1.0643
	Managing Director, Board Member..	.26942	.17938	.894	-.3690	.9078
	Other	.43772	.19666	.550	-.2622	1.1376
Division Manager etc.	Administrator, Team Member..	.08792	.21263	1.000	-.6688	.8447
	Manager without a team	.21120	.19381	.977	-.4786	.9010
	Team Lead, Group Lead	.38549	.21116	.766	-.3660	1.1370
	Department Head	-.32489	.20776	.874	-1.0643	.4145
	Managing Director, Board Member..	-.05547	.17583	1.000	-.6812	.5703
	Other	.11282	.19342	.999	-.5756	.8012
Managing Director, Board Member...	Administrator, Team Member, Assistant	.14339	.18500	.996	-.5150	.8018
	Manager without a team	.26667	.16302	.848	-.3135	.8469
	Team Lead, Group Lead	.44096	.18331	.448	-.2114	1.0934
	Department Head	-.26942	.17938	.894	-.9078	.3690
	Division Manager etc.	.05547	.17583	1.000	-.5703	.6812
	Other	.16829	.16256	.983	-.4103	.7469
Other	Administrator, Team Member..	-.02491	.20179	1.000	-.7431	.6933
	Manager without a team	.09838	.18185	1.000	-.5488	.7456
	Team Lead, Group Lead	.27266	.20025	.932	-.4400	.9853
	Department Head	-.43772	.19666	.550	-1.1376	.2622
	Division Manager etc.	-.11282	.19342	.999	-.8012	.5756
	Managing Director etc.	-.16829	.16256	.983	-.7469	.4103

Annex 17. Impact of Organizational Size on the Channel Type

Table 1.1. Ranking of Organizational Size and Channel Type Phone Call

	Phone	N	Mean Rank	Rank Sum
Size	Not quoted	285	382.06	108887.50
	Quoted	448	357.42	160123.50
	Total	733		

Table 1.2. Test Statistics[a] of Organizational Size and Channel Phone Call

	Size
Mann-Whitney-U-Test	59547.500
Wilcoxon-W	160123.500
Z	-1.569
Assymp. Sig. (2-sided)	.117

a. Group variable: Phone

Table 2.1. Ranking of Organizational Size and Channel Type Face-to-Face

	F2F	N	Mean Rank	Rank Sum
Size	Not quoted	258	367.51	94817.00
	Quoted	475	366.72	174194.00
	Total	733		

Table 2.2. Test Statistics[a] of Organizational Size and Face-to-Face

	Size
Mann-Whitney-U-Test	61144.000
Wilcoxon-W	174194.000
Z	-.049
Assymp. Sig. (2-sided)	.961

a. Group variable: Face-to-face

Table 3.1. Ranking of Organizational Size and Channel Type Email

	Email	N	Mean Rank	Rank Sum
Size	Not quoted	236	369.13	87114.00
	Quoted	497	365.99	181897.00
	Total	733		

Table 3.2. Test Statistics[a] of Organizational Size and Channel Type Email

	Size
Mann-Whitney-U-Test	58144.000
Wilcoxon-W	181897.000
Z	-.191
Assymp. Sig. (2-sided)	.848

a. Group variable: Email

Table 4.1. Ranking of Organizational Size and Channel Type Chat

	Chat	N	Mean Rank	Rank Sum
Size	Not quoted	600	367.22	220333.00
	Quoted	133	366.00	48678.00
	Total	733		

Table 4.2. Test Statistics[a] of Organizational Size and Channel Type Chat

	Size
Mann-Whitney-U-Test	39767.000
Wilcoxon-W	48678.000
Z	-.062
Assymp. Sig. (2-sided)	.951

a. Group variable: Chat (as part of a communication platform e.g., Skype etc.)

Table 5.1. Ranking of Organizational Size Impact and Messaging

	Messaging	N	Mean Rank	Rank Sum
Size	Not quoted	576	387.00	222910.00
	Quoted	157	293.64	46101.00
	Total	733		

Table 5.2. Test Statistics[a] of Organizational Size Impact and Messaging

	Size
Mann-Whitney-U-Test	33698.000
Wilcoxon-W	46101.000
Z	-5.003
Assymp. Sig. (2-sided)	<.001

a. Group variable: Messaging (e.g., WhatsApp etc.)

Table 6.1. Ranking of Organizational Size and Channel Audio Conferencing

	Audio	N	Mean Rank	Rank Sum
Size	Not quoted	622	356.45	221712.50
	Quoted	111	426.11	47298.50
	Total	733		

Table 6.2. Test Statistics[a] of Organizational Size and Audio Conferencing

	Size
Mann-Whitney-U-Test	27959.500
Wilcoxon-W	221712.500
Z	-3.262
Assymp. Sig. (2-sided)	.001

a. Group variable: Audio Conferencing

Table 7.1. Ranking of Organizational Size and Channel Video Conferencing

	Video	N	Mean Rank	Rank Sum
Size	Not quoted	275	341.05	93790.00
	Quoted	458	382.58	175221.00
	Total	733		

Table 7.2. Test Statistics[a] Organizational Size and Video Conferencing

	Size
Mann-Whitney-U-Test	55840.000
Wilcoxon-W	93790.000
Z	-2.626
Assymp. Sig. (2-sided)	.009

a. Group variable: Video Conferencing

Note: Due to the low selection of “other” channel by survey respondents this category is not analyzed.

Annex 18. Impact of Organizational Size on the Channel Quantity

Table 1. Correlation between Organizational Size and Channel Quantity

Correlation			
			Channel Quantity
Spearman-Rho	Size	Correlation Coefficient	-.013
		Sig. (two-sided)	.731
		N	733

Annex 19. Impact of Negotiation Experience on the Channel Type

Table 1.1. Group Statistics for Negotiation Experience and Phone Call Use

Group Statistics					
	Phone Call	N	Mean	Std. Deviation	Std. Error Mean
Negotiation Experience	Not quoted	285	5.21	32.247	1.910
	Quoted	448	11.22	23.809	1.125

Table 1.2. Assessment if Experience Impacts Phone Negotiations through the Test for Independent Samples

		Levene-Test for Variance Equality		T-Test for Average Mean Equality						
		F	Sig.	T	df	Sig. (2-sided)	Mid. Diff.	Diff. Stand. Error	Lower Value	Upper Value
Channel Quantity	Variance is equal	10.268	.001	-2.898	731	.004	-6.016	2.076	-10.092	-1.941
	Variance is not equal			-2.714	478. 573	.007	-6.016	2.217	-10.372	-1.660

Table 1.3. Effect Size of Independent Samples for the Impact of Negotiation Experience on the Channel Type Phone Call

Independent Samples Effect Sizes

		Standardizer[a]	Point Estimate	95% Confidence Interval Lower	Upper
Negotiation Experience	Cohen's d	27.398	-.220	-.368	-.071
	Hedges' Correction	27.426	-.219	-.368	-.071
	Glass' Delta	23.809	-.253	-.402	-.103

a. Common denominator used during the estimation of the effect size. Cohen's d uses the summarized standard deviation. Hedges' Correction uses the summarized standard deviation and a correction factor. Glass' Delta uses the standard deviation of a sample from a control group.

Table 2.1. Group Statistics for Experience and Channel Face-to-Face

Group Statistics

	Face-to-Face	N	Mean	Std. Deviation	Std. Error Mean
Negotiation Experience	Not quoted	258	7.22	29.777	1.854
	Quoted	475	9.79	26.227	1.203

Table 2.2. Assessment if Experience Impacts Face-to-Face Negotiations through the Test for Independent Samples

		Levene-Test for Variance Equality		T-Test for Average Mean Equality						
		F	Sig.	T	df	Sig. (2-sided)	Mid. Diff.	Diff. Stand. Error	Lower Value	Upper Value
Channel Quantity	Variance is equal	2.010	.157	-1.208	731	.227	-2.572	2.129	-6.752	1.607
	Variance is not equal			-1.164	473.622	.245	-2.572	2.210	-6.915	1.771

Table 2.3. Effect Size of Independent Samples for the Impact of Negotiation Experience on the Channel Face-to-Face Used in a Negotiation

Independent Samples Effect Sizes

		Standardizer[a]	Point Estimate	95% Confidence Interval Lower	Upper
Negotiation Experience	Cohen's d	27.527	-.093	-.245	.058
	Hedges' Correction	27.556	-.093	-.245	.058
	Glass' Delta	26.227	-.098	-.250	.054

a. Common denominator used during the estimation of the effect size. Cohen's d uses the summarized standard deviation. Hedges' Correction uses the summarized standard deviation and a correction factor. Glass' Delta uses the standard deviation of a sample from a control group.

Table 3.1. Group Statistics for Experience and Channel Email

Group Statistics

	Email	N	Mean	Std. Deviation	Std. Error Mean
Negotiation Experience	Not quoted	236	3.23	34.702	2.259
	Quoted	497	11.57	22.936	1.029

Table 3.2. Assessment if Experience Impacts the Email through the Test for Independent Samples

		Levene-Test for Variance Equality		T-Test for Average Mean Equality						
		F	Sig.	T	df	Sig. (2-sided)	Mid. Diff.	Diff. Stand. Error	Lower Value	Upper Value
Negotiation Experience	Variance is equal	21.280	<.001	-3.865	731	<.001	-8.334	2.156	-12.568	-4.101
	Variance is not equal			-3.358	335.756	<.001	-8.334	2.482	-13.217	-3.452

Table 3.3. Effect Size of Independent Samples for the Impact of Negotiation Experience on the Channel Email Used in a Negotiation

Independent Samples Effect Sizes

		Standardizer[a]	Point Estimate	95% Confidence Interval Lower	Upper
Negotiation Experience	Cohen's d	27.278	-.306	-.461	-.150
	Hedges' Correction	27.306	-.305	-.461	-.150
	Glass' delta	22.936	-.363	-.520	-.207

a. Common denominator used during the estimation of the effect size. Cohen's d uses the summarized standard deviation. Hedges' Correction uses the summarized standard deviation and a correction factor. Glass' Delta uses the standard deviation of a sample from a control group

Table 4.1. Group Statistics for Experience and Channel Chat

Group Statistics

	Chat	N	Mean	Std. Deviation	Std. Error Mean
Negotiation Experience	Not quoted	600	8.36	28.287	1.155
	Quoted	133	11.24	23.810	2.065

Table 4.2. Assessment if Experience Impacts the Chat Use in a Negotiation through the Test for Independent Samples

		Levene-Test for Variance Equality		T-Test for Average Mean Equality						
		F	Sig.	T	df	Sig. (2-sided)	Mid. Diff.	Diff. Stand. Error	Lower Value	Upper Value
Negotiation Experience	Variance is equal	1.034	.310	-1.091	731	.276	-2.879	2.639	-8.059	2.301
	Variance is not equal			-1.217	222.717	.225	-2.879	2.366	-7.541	1.783

Table 4.3. Effect Size of Independent Samples for the Impact of Negotiation Experience on the Channel Chat Used in a Negotiation

Independent Samples Effect Sizes

		Standardizer[a]	Point Estimate	95% Confidence Interval Lower	Upper
Negotiation Experience	Cohen's d	27.532	-.105	-.292	.083
	Hedges' correction	27.561	-.104	-.292	.083
	Glass' delta	23.810	-.121	-.309	.068

a. Common denominator used during the estimation of the effect size. Cohen's d uses the summarized standard deviation. Hedges' Correction uses the summarized standard deviation and a correction factor. Glass' Delta uses the standard deviation of a sample from a control group.

Table 5.1. Group Statistics for Experience and Channel Messaging

Group Statistics

	Messaging	N	Mean	Std. Deviation	Std. Error Mean
Negotiation Experience	Not quoted	576	8.54	28.110	1.171
	Quoted	157	10.15	25.365	2.024

Table 5.2. Assessment if Experience Impacts the Messaging through the Test for Independent Samples

		Levene-Test for Variance Equality		T-Test for Average Mean Equality						
		F	Sig.	T	df	Sig. (2-sided)	Mid. Diff.	Diff. Stand. Error	Lower Value	Upper Value
Channel Quantity	Variance is equal	.779	.378	-.651	731	.515	-1.615	2.480	-6.484	3.254
	Variance is not equal			-.690	269.724	.491	-1.615	2.339	-6.219	2.990

Table 5.3. Effect Size of Independent Samples for the Impact of Negotiation Experience on the Channel Messaging Used in a Negotiation

Independent Samples Effect Sizes

		Standardizer[a]	Point Estimate	95% Confidence Interval Lower	Upper
Negotiation Experience	Cohen's d	27.547	-.059	-.235	.118
	Hedges' Correction	27.575	-.059	-.235	.118
	Glass' Delta	25.365	-.064	-.240	.113

a. Common denominator used during the estimation of the effect size. Cohen's d uses the summarized standard deviation. Hedges' Correction uses the summarized standard deviation and a correction factor. Glass' Delta uses the standard deviation of a sample from a control group.

Table 6.1. Group Statistics for Experience and Channel Audio Conferencing

Group Statistics

	Audio	N	Mean	Std. Deviation	Std. Error Mean
Negotiation Experience	Not quoted	622	8.77	27.957	1.121
	Quoted	111	9.55	25.152	2.387

Table 6.2. Assessment if Experience Impacts the Audio-Conferencing Use in a Negotiation through the Test for Independent Samples

		Levene-Test for Variance Equality		T-Test for Average Mean Equality						
		F	Sig.	T	df	Sig. (2-sided)	Mid. Diff.	Diff. Stand. Error	Lower Value	Upper Value
Negotiation Experience	Variance is equal	.622	.431	-.276	731	.782	-.784	2.839	-6.358	4.789
	Variance is not equal			-.297	162.456	.767	-.784	2.637	-5.992	4.424

Table 6.3. Effect Size of Independent Samples for the Impact of Negotiation Experience on the Channel Audio Conferencing Used in a Negotiation

		Independent Samples Effect Sizes			
		Standardizer[a]	Point Estimate	95% Confidence Interval Lower	Upper
Negotiation Experience	Cohen's d	27.553	-.028	-.230	.174
	Hedges' Correction	27.582	-.028	-.230	.173
	Glass' Delta	25.152	-.031	-.233	.171

a. Common denominator used during the estimation of the effect size. Cohen's d uses the summarized standard deviation. Hedges' Correction uses the summarized standard deviation and a correction factor. Glass' Delta uses the standard deviation of a sample from a control group.

Table 7.1. Group Statistics for Experience and Channel Video Conferencing

Group Statistics

	Video Conferencing	N	Mean	Std. Deviation	Std. Error Mean
Negotiation Experience	Not quoted	275	4.79	33.493	2.020
	Quoted	458	11.34	22.929	1.071

Table 7.2. Assessment if Experience Impacts the Video Conferencing through the Test for Independent Samples

		Levene-Test for Variance Equality		T-Test for Average Mean Equality						
		F	Sig.	T	df	Sig. (2-sided)	Mid. Diff.	Diff. Stand. Error	Lower Value	Upper Value
Negotiation Experience	Variance is equal	16.337	<.001	-3.141	731	.002	-6.560	2.088	-10.659	-2.460
	Variance is not equal			-2.869	429.518	.004	-6.560	2.286	-11.053	-2.066

Table 7.3. Effect Size of Independent Samples for the Impact of Negotiation Experience on the Channel Video Conferencing Used in a Negotiation

Independent Samples Effect Sizes

		Standardizer[a]	Point Estimate	95% Confidence Interval Lower	Upper
Negotiation Experience	Cohen's d	27.371	-.240	-.390	-.090
	Hedges' Correction	27.399	-.239	-.389	-.089
	Glass' Delta	22.929	-.286	-.437	-.135

a. Common denominator used during the estimation of the effect size. Cohen's d uses the summarized standard deviation. Hedges' Correction uses the summarized standard deviation and a correction factor. Glass' Delta uses the standard deviation of a sample from a control group.

Table 8.1. Group Statistics for Experience and Channel Other

Group Statistics

	Other	N	Mean	Std. Deviation	Std. Error Mean
Negotiation Experience	Not quoted	707	8.81	27.674	1.041
	quoted	26	10.85	23.862	4.680

Table 8.2. Assessment if Experience Impacts the Other through the Test for Independent Samples

		Levene-Test for Variance Equality		T-Test for Average Mean Equality						
		F	Sig.	T	df	Sig. (2-sided)	Mid. Diff.	Diff. Stand. Error	Lower Value	Upper Value
Negotiation Experience	Variance is equal	.312	.577	-.370	731	.712	-2.034	5.502	-12.836	8.767
	Variance is not equal			-.424	27.532	.675	-2.034	4.794	-11.862	7.794

Table 8.3. Effect Size of Independent Samples for the Impact of Negotiation Experience on the Channel Other Used in a Negotiation

Independent Samples Effect Sizes

		Standardizer[a]	Point Estimate	95% Confidence Interval Lower	Upper
Negotiation Experience	Cohen's d	27.552	-.074	-.465	.318
	Hedges' correction	27.581	-.074	-.465	.317
	Glass' delta	23.862	-.085	-.476	.308

a. Common denominator used during the estimation of the effect size. Cohen's d uses the summarized standard deviation. Hedges' Correction uses the summarized standard deviation and a correction factor. Glass' Delta uses the standard deviation of a sample from a control group.

Annex 20. Impact of Negotiation Experience on Channel Quantity

Table 1. Impact of Experience on the Quantity of Channel Selection

	Correlations	Channel Quantity
Kindly let us know your age in the box below.	Pearson Correlation	.166**
	Sig. (two-tailed)	<.001
	N	733

**The correlation is significant on a level of 0.01 (two-tailed).

Annex 21. Impact of History on the Channel Quantity

Table 1. Impact of Negotiation History on the Quantity of Channel Selection

	Correlations	Channel Quantity
Negotiation History	Pearson Correlation	.059
	Sig. (two-tailed)	.109
	N	733

Annex 22. Testing Sychronicity Theory with Channels for Information Sharing and Consensus Building

Table 1.1. Test Statistics of Non-Parametric Friedman-Test for Ranking of Information Sharing Channels in Line with Synchronicity Theory

Test Statistics[a]	
N	733
Chi-Quadrat	244.864
df	6
Asymp. Sig.	0.000

a. Friedman-Test

Table 1.2. Non-Parametric Friedman-Test for Ranking of Consensus Building or Decision-Making Channels in Line with Synchronicity Theory

Test Statistics[a]	
N	733
Chi-Quadrat	277.512
df	6
Asymp. Sig.	0.000

a. Friedman-Test

Annex 23. Gender on Information Sharing and Consensus Building

Table 1.1. Ranks for the Impact of Gender on Information Sharing

Ranks

	Gender	N	Mean Rank	Sum of Ranks
Face-to-Face	Female	179	369.3	66104
	Male	530	350.17	185591
	Total	709		
Phone Call	Female	179	367.11	65712.5
	Male	530	350.91	185982.5
	Total	709		
Audio	Female	179	357.13	63925.5
	Male	530	354.28	187769.5
	Total	709		
Email	Female	179	355.43	63622
	Male	530	354.85	188073
	Total	709		
Messaging	Female	179	365.21	65372.5
	Male	530	351.55	186322.5
	Total	709		
Chat	Female	179	370.08	66244
	Male	530	349.91	185451
	Total	709		
Video	Female	179	349.77	62609.5
	Male	530	356.77	189085.5
	Total	709		

Note: "Other" was also a channel category, however due to the limited cases it is not part of this analysis.

Table 1.2. Ranks for the Impact of Gender on Information Sharing

Test Statistics[a]

	F2F	Phone	Audio	Email	Messaging	Chat	Video
Mann-Whitney U	44876	45267.5	47054.5	47358	45607.5	44736	46499.5
Wilcoxon W	185591	185982.5	187769.5	188073	186322.5	185451	62609.5
Z	-1.157	-0.93	-0.166	-0.033	-0.795	-1.185	-0.406
Asymp. Sig. (2-tailed)	0.247	0.352	0.868	0.974	0.427	0.236	0.685

a. Grouping Variable: Gender

Table 2.1. Ranks for the Impact of Gender on Consensus Building

Ranks				
	Gender	N	Mean Rank	Sum of Ranks
Face-to-Face	Female	179	362.9	64959
	Male	530	352.33	186736
	Total	709		
Phone	Female	179	340.12	60882
	Male	530	360.02	190813
	Total	709		
Audio	Female	179	352.49	63096
	Male	530	355.85	188599
	Total	709		
Email	Female	179	349.57	62573.5
	Male	530	356.83	189121.5
	Total	709		
Messaging	Female	179	360.84	64591
	Male	530	353.03	187104
	Total	709		
Chat	Female	179	348.36	62357
	Male	530	357.24	189338
	Total	709		
Video	Female	179	344.06	61587.5
	Male	530	358.69	190107.5
	Total	709		

Note: “Other” was also a channel category, yet due to the limited cases it is not part of this analysis.

Table 2.2. Ranks for the Impact of Gender on Consensus Building

Test Statistics[a]

	F2F	Phone	Audio	Email	Messaging	Chat	Video
Mann-Whitney U	46021	44772	46986	46463.5	46389	46247	45477.5
Wilcoxon W	186736	60882	63096	62573.5	187104	62357	61587.5
Z	-0.708	-1.148	-0.196	-0.415	-0.46	-0.526	-0.858
Asymp. Sig. (2-tailed)	0.479	0.251	0.845	0.678	0.645	0.599	0.391

a. Grouping Variable: Gender

Annex 24. Impact of Nationality on Channel Type

Table 1.1. Impact of Nationality on the Selection of the Channel Phone Call

Phone Call		Africa (1)	Asia, Southeast Asia & Asia Pacific (2)	Eastern Europe (3)	Middle East (4)	North America (5)	Northern Europe (6)	South America (7)	South Asia (8)	Southern Europe (9)	Western Europe (10)	Total
Not quoted	Quantity	10	15	14	12	27	16	8	15	8	108	233
	Expected Quantity	14.1	11.1	12.6	8.9	21.5	20.4	7.4	19.3	8.1	109.6	233.0
	Standardized Residual	-1.1	1.2	.4	1.0	1.2	-1.0	.2	-1.0	-.1	-.2	
Quoted	Quantity	28	15	20	12	31	39	12	37	14	188	396
	Expected Quantity	23.9	18.9	21.4	15.1	36.5	34.6	12.6	32.7	13.9	186.4	396.0
	Standardized Residual	.8	-.9	-.3	-.8	-.9	.7	-.2	.7	.0	.1	
Total	Quantity	38	30	34	24	58	55	20	52	22	296	629
	Expected Quantity	38.0	30.0	34.0	24.0	58.0	55.0	20.0	52.0	22.0	296.0	629.0

Table 1.2. Chi-Square Tests for the Influence of Nationality on the Selection of the Negotiation Channel Phone Call

	Value	df	Asymptotic Significance (two-sided)
Pearson-Chi-Square	11.368[a]	9	.251
Likelihood-Ratio	11.375	9	.251
Correlation linear-with-linear	.495	1	.482
Quantity of valid cases	629		

a. 0 cells (0.0%) display an expected frequency smaller 5. The minimal expected frequency is 7.41.

Table 1.3. Symmetric Measures for Impact of Nationality on the Selection of the Channel Phone Call for Negotiations

		Value	Approximation Significance
Nominal- regarding Nominal Degree	Phi	.134	.251
	Cramer-V	.134	.251
Quantity of valid cases		629	

Table 2.1. Impact of Nationality on the Selection of the Channel Face-to-Face for Negotiations

Face-to-Face		Africa (1)	Asia, Southeast Asia & Asia Pacific (2)	Eastern Europe (3)	Middle East (4)	North America (5)	Northern Europe (6)	South America (7)	South Asia (8)	Southern Europe (9)	Western Europe (10)	Total
Not quoted	Quantity	10	9	11	10	35	20	7	16	7	96	221
	Expected Quantity	13.4	10.5	11.9	8.4	20.4	19.3	7.0	18.3	7.7	104.0	221.0
	Standardized Residual	-.9	-.5	-.3	.5	3.2	.2	.0	-.5	-.3	-.8	
Quoted	Quantity	28	21	23	14	23	35	13	36	15	200	408
	Expected Quantity	24.6	19.5	22.1	15.6	37.6	35.7	13.0	33.7	14.3	192.0	408.0
	Standardized Residual	.7	.3	.2	-.4	-2.4	-.1	.0	.4	.2	.6	
Total	Quantity	38	30	34	24	58	55	20	52	22	296	629
	Expected Quantity	38.0	30.0	34.0	24.0	58.0	55.0	20.0	52.0	22.0	296.0	629.0

Table 2.2. Chi-Square Tests for the Influence of Nationality on the Selection of the Negotiation Channel Face-to-Face

	Value	df	Asymptotic Significance (two-sided)
Pearson-Chi-Square	19.909[a]	9	.018
Likelihood-Ratio	19.044	9	.025
Correlation linear-with-linear	.704	1	.401
Quantity of valid cases	629		

a. 0 cells (0.0%) display an expected frequency smaller 5. The minimal expected frequency is 7.03.

Table 2.3. Symmetric Measures for Impact of Nationality on the Selection of the Channel Face-to-Face for Negotiations

		Value	Approximation Significance
Nominal- regarding Nominal Degree	Phi	.178	.018
	Cramer-V	.178	.018
Quantity of valid cases		629	

Table 3.1. Impact of Nationality on the Selection of the Channel Email for Negotiations

Email		Africa (1)	Asia, Southeast Asia & Asia Pacific (2)	Eastern Europe (3)	Middle East (4)	North America (5)	Northern Europe (6)	South America (7)	South Asia (8)	Southern Europe (9)	Western Europe (10)	Total
Not quoted	Quantity	21	8	10	8	14	13	6	22	8	84	194
	Expected Quantity	11.7	9.3	10.5	7.4	17.9	17.0	6.2	16.0	6.8	91.3	194.0
	Standardized Residual	2.7	-.4	-.2	.2	-.9	-1.0	-.1	1.5	.5	-.8	
Quoted	Quantity	17	22	24	16	44	42	14	30	14	212	435
	Expected Quantity	26.3	20.7	23.5	16.6	40.1	38.0	13.8	36.0	15.2	204.7	435.0
	Standardized Residual	-1.8	.3	.1	-.1	.6	.6	.0	-1.0	-.3	.5	
Total	Quantity	38	30	34	24	58	55	20	52	22	296	629
	Expected Quantity	38.0	30.0	34.0	24.0	58.0	55.0	20.0	52.0	22.0	296.0	629.0

Table 3.2. Chi-Square Tests for the Influence of Nationality on the Selection of the Negotiation Channel Email

	Value	df	Asymptotic Significance (two-sided)
Pearson-Chi-Square	17.902[a]	9	.036
Likelihood-Ratio	16.925	9	.050
Correlation linear-with-linear	2.069	1	.150
Quantity of valid cases	629		

a. 0 cells (0.0%) display an expected frequency smaller 5. The minimal expected frequency is 6.17.

Table 3.3. Symmetric Measures for Impact of Nationality on the Selection of the Channel Email for Negotiations

		Value	Approximation Significance
Nominal- regarding Nominal Degree	Phi	.169	.036
	Cramer-V	.169	.036
Quantity of valid cases		629	

Table 4.1. Impact of Nationality on the Selection of the Channel Chat for Negotiations

Chat		Africa (1)	Asia, Southeast Asia & Asia Pacific (2)	Eastern Europe (3)	Middle East (4)	North America (5)	Northern Europe (6)	South America (7)	South Asia (8)	Southern Europe (9)	Western Europe (10)	Total
Not quoted	Quantity	28	26	25	17	48	43	14	48	15	248	512
	Expected Quantity	30.9	24.4	27.7	19.5	47.2	44.8	16.3	42.3	17.9	240.9	512.0
	Standardized Residual	-.5	.3	-.5	-.6	.1	-.3	-.6	.9	-.7	.5	
Quoted	Quantity	10	4	9	7	10	12	6	4	7	48	117
	Expected Quantity	7.1	5.6	6.3	4.5	10.8	10.2	3.7	9.7	4.1	55.1	117.0
	Standardized Residual	1.1	-.7	1.1	1.2	-.2	.6	1.2	-1.8	1.4	-1.0	
Total	Quantity	38	30	34	24	58	55	20	52	22	296	629
	Expected Quantity	38.0	30.0	34.0	24.0	58.0	55.0	20.0	52.0	22.0	296.0	629.0

Table 4.2. Chi-Square Tests for the Influence of Nationality on the Selection of the Negotiation Channel Chat

	Value	df	Asymptotic Significance (two-sided)
Pearson-Chi-Square	15.104[a]	9	.088
Likelihood-Ratio	15.102	9	.088
Correlation linear-with-linear	2.714	1	.099
Quantity of valid cases	629		

a. 3 cells (15.0%) display an expected frequency smaller 5. The minimal expected frequency is 3.72.

Table 4.3. Symmetric Measures for Impact of Nationality on the Selection of the Channel Chat for Negotiations

		Value	Approximation Significance
Nominal- regarding Nominal Degree	Phi	.155	.088
	Cramer-V	.155	.088
Quantity of valid cases		629	

Table 5.1. Impact of Nationality on the Selection of the Channel Messaging

Messaging		Africa (1)	Asia, Southeast Asia & Asia Pacific (2)	Eastern Europe (3)	Middle East (4)	North America (5)	Northern Europe (6)	South America (7)	South Asia (8)	Southern Europe (9)	Western Europe (10)	Total
Not quoted	Quantity	21	20	24	14	52	42	8	42	18	250	491
	Expected Quantity	29.7	23.4	26.5	18.7	45.3	42.9	15.6	40.6	17.2	231.1	491.0
	Standardized Residual	-1.6	-.7	-.5	-1.1	1.0	-.1	-1.9	.2	.2	1.2	
Quoted	Quantity	17	10	10	10	6	13	12	10	4	46	138
	Expected Quantity	8.3	6.6	7.5	5.3	12.7	12.1	4.4	11.4	4.8	64.9	138.0
	Standardized Residual	3.0	1.3	.9	2.1	-1.9	.3	3.6	-.4	-.4	-2.4	
Total	Quantity	38	30	34	24	58	55	20	52	22	296	629
	Expected Quantity	38.0	30.0	34.0	24.0	58.0	55.0	20.0	52.0	22.0	296.0	629.0

Table 5.2. Chi-Square Tests for the Influence of Nationality on the Selection of the Negotiation Channel Messaging

	Value	df	Asymptotic Significance (two-sided)
Pearson-Chi-Square	49.411[a]	9	<.001
Likelihood-Ratio	44.485	9	<.001
Correlation linear-with-linear	19.632	1	<.001
Quantity of valid cases	629		

a. 2 cells (0.0%) display an expected frequency smaller 5. The minimal expected frequency is 4.39.

Table 5.3. Symmetric Measures for Impact of Nationality on the Selection of the Channel Messaging for Negotiations

		Value	Approximation Significance
Nominal- regarding Nominal Degree	Phi	.280	<.001
	Cramer-V	.280	<.001
Quantity of valid cases		629	

Table 6.1. Impact of Nationality on the Selection of the Channel Audio Conferencing for Negotiations

Audio		Africa (1)	Asia, Southeast Asia & Asia Pacific (2)	Eastern Europe (3)	Middle East (4)	North America (5)	Northern Europe (6)	South America (7)	South Asia (8)	Southern Europe (9)	Western Europe (10)	Total
Not quoted	Quantity	29	22	30	19	47	51	16	46	22	251	533
	Expected Quantity	32.2	25.4	28.8	20.3	49.1	46.6	16.9	44.1	18.6	250.8	533.0
	Standardized Residual	-.6	-.7	.2	-.3	-.3	.6	-.2	-.3	.8	.0	
Quoted	Quantity	9	8	4	5	11	4	4	6	0	45	96
	Expected Quantity	23.9	18.9	21.4	15.1	36.5	34.6	12.6	32.7	13.9	45.2	96.0
	Standardized Residual	1.3	1.6	-.5	.7	.7	-1.5	.5	-.7	-1.8	.0	
Total	Quantity	38	30	34	24	58	55	20	52	22	296	629
	Expected Quantity	38.0	30.0	34.0	24.0	58.0	55.0	20.0	52.0	22.0	296.0	629.0

Table 6.2. Chi-Square Tests for the Influence of Nationality on the Selection of the Negotiation Channel Audio Conferencing

	Value	df	Asymptotic Significance (two-sided)
Pearson-Chi-Square	14.196[a]	9	.116
Likelihood-Ratio	17.327	9	.044
Correlation linear-with-linear	2.614	1	.106
Quantity of valid cases	629		

a. 4 cells (0.0%) display an expected frequency smaller 5. The minimal expected frequency is 3.05.

Table 6.3. Symmetric Measures for Impact of Nationality on the Selection of the Channel Audio Conferencing for Negotiations

		Value	Approximation Significance
Nominal- regarding Nominal Degree	Phi	.150	.116
	Cramer-V	.150	.116
Quantity of valid cases		629	

Table 7.1. Impact of Nationality on the Selection of the Channel Video Conferencing for Negotiations

Video		Africa (1)	Asia, Southeast Asia & Asia Pacific (2)	Eastern Europe (3)	Middle East (4)	North America (5)	Northern Europe (6)	South America (7)	South Asia (8)	Southern Europe (9)	Western Europe (10)	Total
Not quoted	Quantity	25	14	13	14	18	25	6	25	12	78	230
	Expected Quantity	13.9	11.0	12.4	8.8	21.2	20.1	7.3	19.0	8.0	108.2	230.0
	Standardized Residual	3.0	.9	.2	1.8	-.7	1.1	-.52	1.4	1.4	-2.9	
Quoted	Quantity	13	16	21	10	40	30	14	27	10	218	399
	Expected Quantity	24.1	19.0	21.6	15.2	36.8	34.9	12.7	33.0	14.0	187.8	399.0
	Standardized Residual	-2.3	-.7	-.1	-1.3	.5	-.8	.4	-1.0	-1.1	2.2	
Total	Quantity	38	30	34	24	58	55	20	52	22	296	629
	Expected Quantity	38.0	30.0	34.0	24.0	58.0	55.0	20.0	52.0	22.0	296.0	629.0

Table 7.2. Chi-Square Tests for the Influence of Nationality on the Selection of the Negotiation Channel Phone Call

	Value	df	Asymptotic Significance (two-sided)
Pearson-Chi-Square	42.616[a]	9	<.001
Likelihood-Ratio	42.088	9	<.001
Correlation linear-with-linear	22.393	1	<.001
Quantity of valid cases	629		

a. 0 cells (0.0%) display an expected frequency smaller 5. The minimal expected frequency is 7.31.

Table 7.3. Symmetric Measures for Impact of Nationality on the Selection of the Channel Phone Call for Negotiations

		Value	Approximation Significance
Nominal- regarding Nominal Degree	Phi	.260	<.001
	Cramer-V	.260	<.001
Quantity of valid cases		629	

Note: Due to the low selection of “other” channel by survey respondents this category is not analyzed.

Annex 25. Impact of Nationality on Channel Quantity

Table 1.1. Impact of Nationality on the Channel Quantity

Descriptive Statistics

Channel Quantity					95% Confidence Interval for Mean			
	N	Mean	Std. Deviation	Std. Error	Lower Bound	Upper Bound	Min.	Max.
Africa	38	3.2368	1.60125	0.25976	2.7105	3.7632	1.00	7.00
Asia & Asia Pacific	83	3.0120	1.17384	0.12885	2.7557	3.2684	1.00	6.00
Europe	397	3.2494	1.32603	0.06655	3.1185	3.3802	1.00	7.00
Middle East	24	3.1667	1.71100	0.34926	2.4442	3.8892	1.00	7.00
North America	59	2.9153	1.17866	0.15345	2.6081	3.2224	1.00	6.00
South America	28	3.6429	1.47106	0.27800	3.0724	4.2133	1.00	6.00
Total	629	3.2003	1.33829	0.05336	3.0955	3.3051	1.00	7.00

Table 1.2. Impact of Nationality on the Channel Quantity

Tests of Homogeneity of Variances

		Levene Statistic	df1	df2	Sig.
Channel Quantity	Based on Mean	2,941	5	623	0,012
	Based on Median	1,813	5	623	0,108
	Based on Median and with adjusted df	1,813	5	585,512	0,108
	Based on trimmed mean	2,652	5	623	0,022

Table 1.3. Impact of Nationality on the Channel Quantity

Robust Tests of Equality of Means

Channel Quantity

	Statistics	df1	df2	Sig.
Welch	1,669	5	94,704	0,150

a. Asymptotically F distributed.

Annex 26. Geographic Location and Selection of Channel Type

Table 1.1. Geographic Location and Selection of Phone Call for Negotiations

Phone Call		Africa	Asia & Asia Pacific	Europe	Global	Middle East	North America	South America	Total
Not quoted	Quantity	8	33	141	3	21	33	10	249
	Expected Quantity	11.9	29.9	143.5	7.7	17.6	27.6	10.7	249.0
	Standardized Residual	-1.1	0.6	-0.2	-1.7	0.8	1.0	-0.2	
Quoted	Quantity	23	45	233	17	25	39	18	400
	Expected Quantity	19.1	48.1	230.5	12.3	28.4	44.4	17.3	400.0
	Standardized Residual	0.9	-0.4	0.2	1.3	-0.6	-0.8	0.2	
Total	Quantity	31	78	374	20	46	72	28	649
	Expected Quantity	31.0	78.0	374.0	20.0	46.0	72.0	28.0	649.0

Table 1.2. Chi-Square Tests for the Influence of Country of Operation on the Selection of the Negotiation Channel Phone Call

	Value	df	Asymptotic Significance (two-sided)
Pearson-Chi-Square	10.082[a]	6	0.121
Likelihood-Ratio	10.818	6	0.094
Correlation linear-with-linear	1.226	1	0.268
Quantity of valid cases	649		

a. 0 cells (.0%) display an expected frequency smaller 5. The minimal expected frequency is 7.67.

Table 1.3. Symmetric Measures for Influence of Country on Phone

		Value	Approximation Significance
Nominal- regarding	Phi	0.125	0.121
Nominal Degree	Cramer-V	0.125	0.121
Quantity of valid cases		649	

Table 2.1. Geographic Location and Selection of Face-to-Face for Negotiations

F2F		Africa	Asia & Asia Pacific	Europe	Global	Middle East	North America	South America	Total
Not quoted	Quantity	10	22	121	6	13	42	10	224
	Expected Quantity	10.7	26.9	129.1	6.9	15.9	24.9	9.7	224.0
	Standardized Residual	-0.2	-0.9	-0.7	-0.3	-0.7	3.4	0.1	
Quoted	Quantity	21	56	253	14	33	30	18	425
	Expected Quantity	20.3	51.1	244.9	13.1	30.1	47.1	18.3	425.0
	Standardized Residual	0.2	0.7	0.5	0.2	0.5	-2.5	-0.1	
Total	Quantity	31	78	374	20	46	72	28	649
	Expected Quantity	31.0	78.0	374.0	20.0	46.0	72.0	28.0	649.0

Table 2.2. Chi-Square Tests for the Influence of Country of Operation on the Selection of the Negotiation Channel Face-to-Face

	Value	df	Asymptotic Significance (two-sided)
Pearson-Chi-Square	21.284[a]	6	0.002
Likelihood-Ratio	20.251	6	0.002
Correlation linear-with-linear	8.916	1	0.003
Quantity of valid cases	649		

a. 0 cells (.0%) display an expected frequency smaller 5. The minimal expected frequency is 6.90.

Table 2.3. Symmetric Measures for Influence of Country of Operation on Face-to-Face in Negotiations

		Value	Approximation Significance
Nominal- regarding Nominal Degree	Phi	0.181	0.002
	Cramer-V	0.181	0.002
Quantity of valid cases		649	

Table 3.1. Geographic Location and Selection of Email for Negotiations

Email		Africa	Asia & Asia Pacific	Europe	Global	Middle East	North America	South America	Total
Not quoted	Quantity	17	26	109	5	18	16	10	201
	Expected Quantity	9.6	24.2	115.8	6.2	14.2	22.3	8.7	201.0
	Standardized Residual	2.4	0.4	-0.6	-0.5	1.0	-1.3	0.5	
Quoted	Quantity	14	52	265	15	28	56	18	448
	Expected Quantity	21.4	53.8	258.2	13.8	31.8	49.7	19.3	448.0
	Standardized Residual	-1.6	-0.3	0.4	0.3	-0.7	0.9	-0.3	
Total	Quantity	31	78	374	20	46	72	28	649
	Expected Quantity	31.0	78.0	374.0	20.0	46.0	72.0	28.0	649.0

Table 3.2. Chi-Square Tests for the Influence of Country of Operation on the Selection of the Negotiation Channel Email

	Value	df	Asymptotic Significance (two-sided)
Pearson-Chi-Square	13.686[a]	6	0.033
Likelihood-Ratio	13.087	6	0.042
Correlation linear-with-linear	1.959	1	0.162
Quantity of valid cases	649		

a. 0 cells (.0%) display an expected frequency smaller 5. The minimal expected frequency is 6.18.

Table 3.3. Symmetric Measures for Influence of Country of Operation on Email in Negotiations

		Value	Approximation Significance
Nominal- regarding Nominal Degree	Phi	0.145	0.033
	Cramer-V	0.145	0.033
Quantity of valid cases		649	

Table 4.1. Geographic Location and Selection of Chat for Negotiations

Chat		Africa	Asia & Asia Pacific	Europe	Global	Middle East	North America	South America	Total
Not quoted	Quantity	21	68	304	17	39	60	21	530
	Expected Quantity	25.3	63.7	305.4	16.3	37.6	58.8	22.9	530.0
	Standardized Residual	-0.9	0.5	-0.1	0.2	0.2	0.2	-0.4	
Quoted	Quantity	10	10	70	3	7	12	7	119
	Expected Quantity	5.7	14.3	68.6	3.7	8.4	13.2	5.1	119.0
	Standardized Residual	1.8	-1.1	0.2	-0.3	-0.5	-0.3	0.8	
Total	Quantity	31	78	374	20	46	72	28	649

Table 4.2. Chi-Square Tests for the Influence of Country of Operation on the Selection of the Negotiation Channel Messaging

	Value	df	Asymptotic Significance (two-sided)
Pearson-Chi-Square	7.045[a]	6	0.317
Likelihood-Ratio	6.588	6	0.361
Correlation linear-with-linear	0.079	1	0.779
Quantity of valid cases	649		

a. 1 cell (7.1%) display an expected frequency smaller 5. The minimal expected frequency is 3.67.

Table 4.3. Symmetric Measures for Influence of Country of Operation on Chat in Negotiations

		Value	Approximation Significance
Nominal- regarding Nominal Degree	Phi	0.104	0.317
	Cramer-V	0.104	0.317
Quantity of valid cases		649	

Table 5.1. Geographic Location and Selection of Messaging for Negotiations

Messaging		Africa	Asia & Asia Pacific	Europe	Global	Middle East	North America	South America	Total
Not quoted	Quantity	17	57	312	17	34	64	12	513
	Expected Quantity	24.5	61.7	295.6	15.8	36.4	56.9	22.1	513.0
	Standardized Residual	-1.5	-0.6	1.0	0.3	-0.4	0.9	-2.2	
Quoted	Quantity	14	21	62	3	12	8	16	136
	Expected Quantity	6.5	16.3	78.4	4.2	9.6	15.1	5.9	136.0
	Standardized Residual	2.9	1.2	-1.8	-0.6	0.8	-1.8	4.2	
Total	Quantity	31	78	374	20	46	72	28	649

Table 5.2. Chi-Square Tests for the Influence of Country of Operation on the Selection of the Negotiation Channel Messaging

	Value	df	Asymptotic Significance (two-sided)
Pearson-Chi-Square	44.479[a]	6	0.000
Likelihood-Ratio	38.661	6	0.000
Correlation linear-with-linear	0.043	1	0.835
Quantity of valid cases	649		

a. 1 cells (7.1%) display an expected frequency smaller 5. The minimal expected frequency is 4.19.

Table 5.3. Symmetric Measures for Influence of Country of Operation on Messaging in Negotiations

		Value	Approximation Significance
Nominal- regarding Nominal Degree	Phi	0.262	0.000
	Cramer-V	0.262	0.000
Quantity of valid cases		649	

Table 6.1. Geographic Location and Selection of Audio Conferencing for Negotiations

Audio		Africa	Asia & Asia Pacific	Europe	Global	Middle East	North America	South America	Total
Not quoted	Quantity	22	62	330	16	40	61	22	553
	Expected Quantity	26.4	66.5	318.7	17.0	39.2	61.3	23.9	553.0
	Standardized Residual	-0.9	-0.5	0.6	-0.3	0.1	0.0	-0.4	
Quoted	Quantity	9	16	44	4	6	11	6	96
	Expected Quantity	4.6	11.5	55.3	3.0	6.8	10.7	4.1	96.0
	Standardized Residual	2.1	1.3	-1.5	0.6	-0.3	0.1	0.9	
Total	Quantity	31	78	374	20	46	72	28	649

Table 6.2. Chi-Square Tests for the Influence of Country of Operation on the Selection of the Negotiation Channel Audio Conferencing

	Value	df	Asymptotic Significance (two-sided)
Pearson-Chi-Square	11.266[a]	6	0.080
Likelihood-Ratio	10.231	6	0.115
Correlation linear-with-linear	0.168	1	0.682
Quantity of valid cases	649		

a. 3 cells (21.4%) display an expected frequency smaller 5. The minimal expected frequency is 2.96.

Table 6.3. Symmetric Measures for Influence of Country of Operation on Audio Conferencing in Negotiations

		Value	Approximation Significance
Nominal- regarding Nominal Degree	Phi	0.132	0.080
	Cramer-V	0.132	0.080
Quantity of valid cases		649	

Table 7.1. Geographic Location and Selection of Video Conferencing for Negotiations

Video		Africa	Asia & Asia Pacific	Europe	Global	Middle East	North America	South America	Total
Not quoted	Quantity	19	35	117	6	21	22	9	229
	Expected Quantity	10.9	27.5	132.0	7.1	16.2	25.4	9.9	229.0
	Standardized Residual	2.4	1.4	-1.3	-0.4	1.2	-0.7	-0.3	
Quoted	Quantity	12	43	257	14	25	50	19	420
	Expected Quantity	20.1	50.5	242.0	12.9	29.8	46.6	18.1	420.0
	Standardized Residual	-1.8	-1.1	1.0	0.3	-0.9	0.5	0.2	
Total	Quantity	31	78	374	20	46	72	28	649

Table 7.2. Chi-Square Tests for the Influence of Country of Operation on the Selection of the Negotiation Channel Video Conferencing

	Value	df	Asymptotic Significance (two-sided)
Pearson-Chi-Square	18.179[a]	6	0.006
Likelihood-Ratio	17.541	6	0.007
Correlation linear-with-linear	2.670	1	0.102
Quantity of valid cases	649		

a. 0 cells (0.0%) display an expected frequency smaller 5. The minimal expected frequency is 7.06.

Table 7.3. Symmetric Measures for Influence of Country of Operation on Video Conferencing in Negotiations

		Value	Approximation Significance
Nominal- regarding Nominal Degree	Phi	0.167	0.006
	Cramer-V	0.167	0.006
Quantity of valid cases		649	

Note: Due to the low selection of "other" channel by survey respondents this category is not analyzed.

Annex 27. Impact of Geographic Location on Channel Quantity

Table 1.1. Descriptive Statistics

Descriptive Statistics

Channel Quantity					95% Confidence Interval Mean Value			
	Sum	Mean	Std.-Deviation	Std.-Error	Lower Limit	Upper Limit	Min.	Max.
Africa	31	3.3226	1.57876	0.28355	2.7435	3.9017	1.00	7.00
Asia & Asia Pacific	78	3.1795	1.29679	0.14683	2.8871	3.4719	1.00	6.00
Europe	374	3.1979	1.32754	0.06865	3.0629	3.3328	1.00	7.00
Global	20	3.5500	1.43178	0.32016	2.8799	4.2201	1.00	6.00
Middle East	46	2.9783	1.37419	0.20261	2.5702	3.3863	1.00	7.00
North America	72	2.9167	1.14756	0.13524	2.6470	3.1863	1.00	6.00
South America	28	3.7143	1.48716	0.28105	3.1376	4.2909	1.00	6.00
Total	649	3.1880	1.33626	0.05245	3.0850	3.2910	1.00	7.00

Table 1.2. Homogeneity Variance Test Geography and Channel Quantity

Tests of Homogeneity of Variances

		Levene Statistic	df1	df2	Sig.
Channel Quantity	Based on Mean	1.329	6	642	0.242
	Based on Median	1.021	6	642	0.410
	Based on Median and with adjusted df	1.021	6	619.982	0.410
	Based on trimmed mean	1.273	6	642	0.268

Table 1.3. ANOVA for Geographic Location and Channel Quantity

ANOVA

		Sum of Squares	df	Mean Square	F	Sig.
Channel Quantity	Between Groups	18.304	6	3.051	1.720	0.114
	Within Groups	1138.762	642	1.774		
	Total	1157.066	648			

Annex 28. Path Coefficients of Direct Relationships

	Path Coefficients
Agility -> Effectiveness	0.108
Agility -> Efficiency	0.188
Behavioral Control -> Effectiveness	0.157
Behavioral Control -> Efficiency	0.207
Cognitive Resource Demands -> Effectiveness	0.086
Cognitive Resource Demands -> Efficiency	0.051
Complexity -> Effectiveness	0.138
Complexity -> Efficiency	0.107
Performance Risk -> Effectiveness	-0.051
Performance Risk -> Efficiency	-0.056
Self-Efficacy -> Effectiveness	0.181
Self-Efficacy -> Efficiency	0.125
Synchronicity -> Effectiveness	0.061
Synchronicity -> Efficiency	-0.043

Annex 29. Path Coefficients of Indirect Relationships

	Path Coefficients
Complexity x Self Efficacy -> Effectiveness	-0.040
Complexity x Self Efficacy -> Efficiency	-0.073
Complexity x Synchronicity -> Effectiveness	-0.060
Complexity x Synchronicity -> Efficiency	-0.003
Complexity x Agility -> Effectiveness	-0.043
Complexity x Agility -> Efficiency	-0.039
Complexity x Behavioral Control -> Effectiveness	0.186
Complexity x Behavioral Control -> Efficiency	0.172
Performance Risk x Self Efficacy -> Effectiveness	-0.031
Performance Risk x Self Efficacy -> Efficiency	0.034
Performance Risk x Agility -> Effectiveness	0.030
Performance Risk x Agility -> Efficiency	0.034
Performance Risk x Synchronicity -> Effectiveness	-0.010
Performance Risk x Synchronicity -> Efficiency	-0.131
Performance Risk x Behavioral Control -> Effectiveness	-0.039
Performance Risk x Behavioral Control -> Efficiency	0.059
Cognitive Resource Demands x Self Efficacy -> Effectiveness	-0.039
Cognitive Resource Demands x Self Efficacy -> Efficiency	-0.000
Cognitive Resource Demands x Agility -> Effectiveness	0.021
Cognitive Resource Demands x Agility -> Efficiency	-0.002
Cognitive Resource Demands x Synchronicity -> Effectiveness	-0.099
Cognitive Resource Demands x Synchronicity -> Efficiency	-0.019
Cognitive Resource Demands x Behavioral Control -> Effectiveness	-0.004
Cognitive Resource Demands x Behavioral Control -> Efficiency	-0.058

Annex 30. Quality Criteria

Table 1.1. R-Squared

	R-square	R-square adjusted
Effectiveness	0.290	0.271
Efficiency	0.237	0.216

Table 1.2. F-Square in Direct Causal Variable Relationships

Causal Variable Relationship	F-Squared
Agility -> Effectiveness	0.010
Agility -> Efficiency	0.027
Behavioral Control -> Effectiveness	0.018
Behavioral Control -> Efficiency	0.030
Cognitive Resource Demands -> Effectiveness	0.009
Cognitive Resource Demands -> Efficiency	0.003
Complexity -> Effectiveness	0.018
Complexity -> Efficiency	0.010
Performance Risk -> Effectiveness	0.003
Performance Risk -> Efficiency	0.003
Self Efficacy -> Effectiveness	0.036
Self Efficacy -> Efficiency	0.016
Synchronicity-> Effectiveness	0.003
Synchronicity -> Efficiency	0.001

Table 1.3. F-Squared in Indirect Causal Variable Relationships

Causal Variable Relationship	**F-Squared**
Complexity x Self Efficacy -> Effectiveness	0.002
Complexity x Self Efficacy -> Efficiency	0.006
Complexity x Synchronicity-> Effectiveness	0.003
Complexity x Synchronicity -> Efficiency	0.000
Complexity x Agility -> Effectiveness	0.002
Complexity x Agility -> Efficiency	0.001
Complexity x Behavioral Control -> Effectiveness	0.030
Complexity x Behavioral Control -> Efficiency	0.023
Performance Risk x Self Efficacy -> Effectiveness	0.001
Performance Risk x Self Efficacy -> Efficiency	0.001
Performance Risk x Agility -> Effectiveness	0.001
Performance Risk x Agility -> Efficiency	0.001
Performance Risk x Synchronicity -> Effectiveness	0.000
Performance Risk x Synchronicity -> Efficiency	0.014
Performance Risk x Behavioral Control -> Effectiveness	0.002
Performance Risk x Behavioral Control -> Efficiency	0.003
Cognitive Resource Demands x Self Efficacy -> Effectiveness	0.002
Cognitive Resource Demands x Self Efficacy -> Efficiency	0.000
Cognitive Resource Demands x Agility -> Effectiveness	0.000
Cognitive Resource Demands x Agility -> Efficiency	0.000
Cognitive Resource Demands x Synchronicity -> Effectiveness	0.008
Cognitive Resource Demands x Synchronicity -> Efficiency	0.000
Cognitive Resource Demands x Behavioral Control -> Effectiveness	0.000
Cognitive Resource Demands x Behavioral Control -> Efficiency	0.002

Annex 31. Discriminant Validity

Table 1.1. Heterotrait-Monotrait Ratio in Variable Relationships

Variable Relationship	Heterotrait-Monotrait Ratio (HTMT)
Behavioral Control <-> Agility	0.673
Cognitive Resource Demands <-> Agility	0.117
Cognitive Resource Demands <-> Behavioral Control	0.052
Complexity <-> Agility	0.449
Complexity <-> Behavioral Control	0.466
Complexity <-> Cognitive Resource Demands	0.315
Effectiveness <-> Agility	0.452
Effectiveness <-> Behavioral Control	0.463
Effectiveness <-> Cognitive Resource Demands	0.200
Effectiveness <-> Complexity	0.387
Efficiency <-> Agility	0.481
Efficiency <-> Behavioral Control	0.473
Efficiency <-> Cognitive Resource Demands	0.194
Efficiency <-> Complexity	0.344
Efficiency <-> Effectiveness	0.628
Performance Risk <-> Agility	0.195
Performance Risk <-> Behavioral Control	0.240
Performance Risk <-> Cognitive Resource Demands	0.049
Performance Risk <-> Complexity	0.332
Performance Risk <-> Effectiveness	0.155
Performance Risk <-> Efficiency	0.142
Self-Efficacy <-> Agility	0.334
Self-Efficacy <-> Behavioral Control	0.409
Self-Efficacy <-> Cognitive Resource Demands	0.211
Self-Efficacy <-> Complexity	0.320
Self-Efficacy <-> Effectiveness	0.435
Self-Efficacy <-> Efficiency	0.343
Self-Efficacy <-> Performance Risk	0.214
Synchronicity <-> Agility	0.702
Synchronicity <-> Behavioral Control	0.692
Synchronicity <-> Cognitive Resource Demands	0.168
Synchronicity <-> Complexity	0.479
Synchronicity <-> Effectiveness	0.436
Synchronicity <-> Efficiency	0.331
Synchronicity <-> Performance Risk	0.267
Synchronicity <-> Self-Efficacy	0.366

Table 1.2. Fornell-Larcker Criterion

	A	BC	CRD	C	EFFE	EFFI	PR	SE	S
Agility	0.771								
Behavioral Control	0.535	0.806							
Cognitive Resource Demands	0.070	0.044	0.878						
Complexity	0.367	0.397	0.257	0.604					
Effectiveness	0.357	0.379	0.173	0.319	0.743				
Efficiency	0.353	0.356	0.106	0.257	0.470	0.657			
Performance Risk	-0.153	-0.199	0.032	0.125	-0.113	-0.110	0.851		
Self-Efficacy	0.270	0.342	0.166	0.246	0.335	0.274	-0.115	0.780	
Synchronicity	0.553	0.568	0.149	0.421	0.354	0.253	-0.217	0.287	0.855

Table 1.3. Cross Loadings Fornell-Larcker Criterion Direct Relationships

	A	BC	CRD	C	EFFE	EFFI	PR	SE	SY
Agility 1	0.785	0.405	0.083	0.311	0.315	0.263	-0.141	0.252	0.471
Agility 2	0.836	0.440	-0.021	0.260	0.280	0.327	-0.118	0.169	0.391
Agility 3	0.733	0.398	0.045	0.259	0.273	0.249	-0.079	0.208	0.381
Agility 4	0.723	0.409	0.126	0.308	0.225	0.241	-0.136	0.209	0.473
Behavioral Control 1	0.416	0.795	0.015	0.317	0.371	0.261	-0.193	0.392	0.446
Behavioral Control 2	0.430	0.789	0.040	0.309	0.256	0.283	-0.084	0.230	0.425
Behavioral Control 3	0.412	0.830	0.053	0.304	0.280	0.294	-0.159	0.257	0.455
Behavioral Control 4	0.468	0.812	0.036	0.348	0.303	0.311	-0.196	0.213	0.501
Cogn. Res. Dem. 1	-0.017	0.001	0.815	0.194	0.125	0.061	0.042	0.087	0.089
Cogn. Res. Dem. 2	0.108	0.068	0.936	0.254	0.183	0.137	0.006	0.177	0.173
Cogn. Res. Dem. 3	0.064	0.029	0.878	0.220	0.133	0.059	0.050	0.159	0.109
Complexity 1	0.368	0.385	0.202	0.649	0.264	0.201	-0.145	0.252	0.459
Complexity 2	0.270	0.276	0.209	0.661	0.200	0.136	0.062	0.138	0.307
Complexity 3	0.166	0.197	0.195	0.643	0.197	0.120	0.088	0.180	0.298
Complexity 4	0.233	0.285	0.102	0.630	0.261	0.192	0.041	0.186	0.261
Complexity 5	0.093	0.073	0.149	0.490	0.106	0.124	0.256	0.071	0.089
Complexity 6	0.161	0.153	0.130	0.578	0.115	0.153	0.242	0.071	0.088
Complexity 7	0.131	0.160	0.088	0.558	0.098	0.124	0.248	0.019	0.071
Effectiveness 1	0.299	0.314	0.170	0.269	0.833	0.393	-0.109	0.290	0.310
Effectiveness 2	0.268	0.332	0.131	0.210	0.814	0.406	-0.120	0.290	0.236
Effectiveness 3	0.291	0.326	0.105	0.233	0.830	0.378	-0.113	0.252	0.275
Effectiveness 4	0.277	0.236	0.150	0.286	0.700	0.309	-0.053	0.239	0.317
Effectiveness 5	0.169	0.170	0.069	0.185	0.476	0.230	0.013	0.143	0.144
Efficiency 1	0.162	0.171	0.078	0.093	0.170	0.549	-0.031	0.087	0.112
Efficiency 2	0.168	0.254	0.164	0.238	0.274	0.660	-0.078	0.226	0.169
Efficiency 3	0.259	0.248	0.149	0.182	0.377	0.769	-0.093	0.255	0.189
Efficiency 4	0.258	0.219	-0.057	0.141	0.350	0.670	-0.054	0.131	0.113
Efficiency 5	0.296	0.260	-0.004	0.165	0.331	0.618	-0.086	0.159	0.225
Performance Risk 1	-0.158	-0.195	0.020	0.108	-0.102	-0.087	0.845	-0.117	-0.236
Performance Risk 2	-0.082	-0.126	0.060	0.158	-0.078	-0.112	0.835	-0.058	-0.084
Performance Risk 3	-0.152	-0.188	0.001	0.052	-0.108	-0.081	0.872	-0.119	-0.233
Self-Efficacy 1	0.276	0.347	0.131	0.229	0.303	0.241	-0.104	0.902	0.279
Self-Efficacy 2	0.238	0.301	0.166	0.249	0.309	0.282	-0.067	0.922	0.250
Self-Efficacy 3_(R)	0.035	0.054	0.088	-0.010	0.125	0.020	-0.192	0.400	0.098
Synchronicity 1	0.433	0.442	0.123	0.359	0.307	0.207	-0.148	0.241	0.830
Synchronicity 2	0.492	0.506	0.126	0.353	0.301	0.245	-0.189	0.243	0.879
Synchronicity 3	0.493	0.509	0.134	0.369	0.299	0.194	-0.221	0.252	0.856

Table 1.4. Cross Loadings Fornell-Larcker Criterion Indirect Relationships 1/2

	A	BC	CRD	C	EFFE	EFFI	PR	SE	SY
Complexity x Agility	-0.217	-0.132	-0.004	-0.167	-0.119	-0.050	-0.051	-0.132	-0.244
Cognitive Resource Demands x Agility	-0.062	0.041	-0.027	-0.006	-0.074	-0.030	-0.042	0.026	-0.063
Complexity x Behavioral Control	-0.139	-0.147	-0.026	-0.207	-0.036	0.022	-0.038	-0.164	-0.209
Cognitive Resource Demands x Synchronicity	-0.061	-0.008	-0.047	-0.034	-0.145	-0.058	-0.008	0.006	-0.079
Performance Risk x Behavioral Control	0.009	0.063	-0.044	-0.048	-0.044	0.024	0.008	-0.111	0.054
Complexity x Self-Efficacy	-0.136	-0.161	-0.018	-0.193	-0.132	-0.087	-0.062	-0.216	-0.208
Complexity x Synchronicity	-0.239	-0.194	-0.024	-0.213	-0.157	-0.062	-0.035	-0.196	-0.309
Performance Risk x Self-Efficacy	-0.146	-0.126	-0.006	-0.091	-0.117	-0.030	0.071	-0.079	-0.174
Performance Risk x Synchronicity	-0.021	0.056	-0.009	-0.050	-0.041	-0.084	0.017	-0.160	0.074
Cognitive Resource Demands x Self-Efficacy	0.029	0.000	0.060	-0.027	-0.068	-0.029	-0.007	0.013	0.007
Cognitive Resource Demands x Behavioral Control	0.044	0.113	0.003	-0.038	-0.033	-0.007	-0.049	0.000	-0.008
Performance Risk x Agility	0.023	0.009	-0.044	-0.076	-0.043	-0.016	0.064	-0.140	-0.022

Table 1.5. Cross Loadings Fornell-Larcker Criterion Indirect Relationships 2/2

	C x SE	C x S	C x A	C x BC	PR x SE	PR x A	PR x S	PR x BC	CRD x SE	CRD x A	CRD x S	CRD x BC
C x A	0.674	0.848	1.000	0.766	0.277	0.248	0.248	0.233	0.101	0.311	0.225	0.186
CRD x A	0.117	0.241	0.311	0.177	0.030	0.091	0.028	0.112	0.273	1.000	0.658	0.576
C x BC	0.716	0.814	0.766	1.000	0.246	0.259	0.246	0.260	0.126	0.177	0.161	0.256
CRD x S	0.136	0.267	0.225	0.161	0.009	0.034	-0.027	0.064	0.325	0.658	1.000	0.640
PR x BC	0.216	0.218	0.233	0.260	0.326	0.613	0.617	1.000	0.018	0.112	0.064	0.099
C x SE	1.000	0.730	0.674	0.716	0.295	0.272	0.297	0.216	0.250	0.117	0.136	0.150
C x S	0.730	1.000	0.848	0.814	0.302	0.244	0.234	0.218	0.130	0.241	0.267	0.188
PR x SE	0.295	0.302	0.277	0.246	1.000	0.355	0.297	0.326	0.061	0.030	0.009	-0.004
PR x S	0.297	0.234	0.248	0.246	0.297	0.647	1.000	0.617	0.025	0.028	-0.027	0.058
CRD x SE	0.250	0.130	0.101	0.126	0.061	0.048	0.025	0.018	1.000	0.273	0.325	0.436
CRD x BC	0.150	0.188	0.186	0.256	-0.004	0.123	0.058	0.099	0.436	0.576	0.640	1.000
PR x A	0.272	0.244	0.248	0.259	0.355	1.000	0.647	0.613	0.048	0.091	0.034	0.123

Annex 32. Causal Relationships of Behavioral Dimensions

	Original Sample	Sample Mean	STDV	T Statistics	P Values
Self-Efficacy -> Effectiveness	0.185	0.186	0.042	4.372	0.000
Agility -> Efficiency	0.200	0.203	0.049	4.057	0.000
Behavioral Control -> Efficiency	0.186	0.181	0.054	3.453	0.000
Behavioral Control -> Effectiveness	0.161	0.161	0.047	3.414	0.000
Self-Efficacy -> Efficiency	0.128	0.129	0.038	3.332	0.000
Agility -> Effectiveness	0.112	0.109	0.041	2.729	0.003
Synchronicity -> Efficiency	-0.050	-0.052	0.052	0.975	0.165
Synchronicity -> Effectiveness	0.047	0.046	0.049	0.956	0.170

Annex 33. Moderating Effects

Table 1. Causal Relationships of Moderating Effects

	Original Sample	Sample Mean	STDV	T Statistics	P Values
Complexity x Behavioral Control -> Effectiveness	0.182	0.182	0.043	4.230	0.000
Complexity x Behavioral Control -> Efficiency	0.152	0.152	0.045	3.370	0.000
Complexity -> Effectiveness	0.142	0.146	0.044	3.211	0.001
Complexity -> Efficiency	0.127	0.131	0.046	2.765	0.003
Cognitive Resource Demands -> Effectiveness	0.087	0.086	0.037	2.329	0.010
Performance Risk x Synchronicity -> Efficiency	-0.122	-0.125	0.053	2.288	0.011
Cognitive Resource Demands x Synchronicity -> Effectiveness	-0.099	-0.089	0.050	1.984	0.024
Performance Risk -> Efficiency	-0.060	-0.062	0.040	1.479	0.070
Performance Risk -> Effectiveness	-0.049	-0.052	0.038	1.289	0.099
Complexity x Self-Efficacy -> Efficiency	-0.052	-0.051	0.041	1.276	0.101
Performance Risk x Behavioral Control -> Efficiency	0.058	0.057	0.049	1.195	0.116
Complexity x Agility -> Effectiveness	-0.047	-0.041	0.042	1.121	0.131
Complexity x Synchronicity -> Effectiveness	-0.058	-0.060	0.053	1.092	0.137
Performance Risk x Behavioral Control -> Effectiveness	-0.043	-0.038	0.040	1.071	0.142
Performance Risk x Self-Efficacy -> Effectiveness	-0.046	-0.046	0.047	0.970	0.166
Cognitive Resource Demands -> Efficiency	0.040	0.039	0.043	0.922	0.178
Complexity x Self-Efficacy -> Effectiveness	-0.042	-0.043	0.054	0.777	0.219
Cognitive Resource Demands x Behavioral Control -> Efficiency	-0.042	-0.042	0.055	0.763	0.223
Cognitive Resource Demands x Synchronicity -> Efficiency	-0.032	-0.028	0.048	0.679	0.249
Performance Risk x Self-Efficacy -> Efficiency	0.028	0.025	0.042	0.664	0.253
Performance Risk x Agility -> Effectiveness	0.031	0.028	0.047	0.658	0.255
Cognitive Resource Demands x Self-Efficacy -> Effectiveness	-0.032	-0.025	0.052	0.616	0.269
Cognitive Resource Demands x Agility -> Effectiveness	0.025	0.026	0.048	0.528	0.299
Complexity x Synchronicity -> Efficiency	-0.027	-0.025	0.053	0.520	0.302
Performance Risk x Agility -> Efficiency	0.026	0.030	0.054	0.481	0.315
Complexity x Agility -> Efficiency	-0.018	-0.017	0.048	0.371	0.355
Cognitive Resource Demands x Behavioral Control -> Effectiveness	-0.015	-0.015	0.052	0.294	0.384
Cognitive Resource Demands x Agility -> Efficiency	0.002	0.007	0.047	0.037	0.485
Cognitive Resource Demands x Self-Efficacy -> Efficiency	0.001	0.009	0.048	0.021	0.492
Performance Risk x Synchronicity -> Effectiveness	0.000	-0.005	0.054	0.007	0.497

Table 1.1. R-Squared Through Moderating Variables – Complexity

	Moderators Included	Moderators Excluded
R-Squared Effectiveness	0.262	0.225
R-Squared Efficiency	0.210	0.184

Table 1.2. R-Squared Through Moderation - Cognitive Demands

	Moderators Included	Moderators Excluded
R-Squared Effectiveness	0.254	0.225
R-Squared Efficiency	0.189	0.184

Table 1.3. R-Squared Through Moderation - Performance Risk

	Moderators Included	Moderators Excluded
R-Squared Effectiveness	0.228	0.225
R-Squared Efficiency	0.196	0.184

Annex 34. F-Squared

Table 1. F-Squared for Independent and Moderating Variables

	Effectiveness	Efficiency
Self-Efficacy	0.038	0.017
Agility	0.010	0.030
Synchronicity	0.002	0.002
Behavioral Control	0.020	0.024
Complexity	0.019	0.014
Performance Risk	0.003	0.004
Cognitive Resource Demands	0.009	0.002
Cognitive Resource Demands x Agility	0.001	0.000
Performance Risk x Behavioral Control	0.002	0.003
Complexity x Synchronicity	0.002	0.000
Cognitive Resource Demands x Self-Efficacy	0.001	0.000
Performance Risk x Synchronicity	0.000	0.012
Complexity x Agility	0.002	0.000
Performance Risk x Agility	0.001	0.000
Complexity x Self-Efficacy	0.002	0.003
Cognitive Resource Demands x Behavioral Control	0.000	0.001
Complexity x Behavioral Control	0.028	0.018
Cognitive Resource Demands x Synchronicity	0.008	0.001
Performance Risk x Self-Efficacy	0.003	0.001

Annex 35. Control Variables

Table 1.1. Complete List of Control Variables 1/2

	Original Sample	Sample Mean	Standard Deviation	T Statistics	P Value
Organization Size -> Effectiveness	0.077	0.076	0.035	2.194	0.014
Divisional Head -> Effectiveness	-0.269	-0.270	0.126	2.133	0.016
Industrial Goods -> Effectiveness	0.262	0.254	0.123	2.132	0.017
Public -> Efficiency	-0.405	-0.397	0.195	2.072	0.019
Media & Telecommunications -> Efficiency	0.361	0.370	0.182	1.990	0.023
Team Lead -> Effectiveness	-0.198	-0.198	0.120	1.645	0.050
Media & Telecommunications -> Effectiveness	0.257	0.265	0.168	1.528	0.063
Chemicals -> Effectiveness	-0.175	-0.176	0.115	1.521	0.064
Management -> Efficiency	0.131	0.132	0.092	1.436	0.076
Negotiation History -> Effectiveness	0.094	0.094	0.069	1.361	0.087
Specialist -> Effectiveness	-0.173	-0.172	0.138	1.253	0.105
Purchasing -> Effectiveness	-0.106	-0.094	0.089	1.195	0.116
Industrial Goods -> Efficiency	-0.163	-0.176	0.143	1.140	0.127
Female -> Efficiency	0.084	0.075	0.077	1.078	0.140
Sales -> Effectiveness	-0.146	-0.149	0.139	1.051	0.147
Purchasing -> Efficiency	-0.103	-0.088	0.100	1.028	0.152
Department Head -> Efficiency	0.133	0.128	0.133	1.001	0.158
Services -> Effectiveness	0.096	0.102	0.100	0.964	0.167
Manager without Team -> Efficiency	0.121	0.117	0.126	0.962	0.168
Other -> Effectiveness	-0.253	-0.255	0.276	0.916	0.180
Manager without Team -> Effectiveness	-0.116	-0.104	0.135	0.856	0.196
Consumer Goods & Retail -> Effectiveness	-0.113	-0.114	0.132	0.856	0.196
Other -> Efficiency	-0.193	-0.183	0.226	0.853	0.197
Department Head -> Effectiveness	-0.098	-0.083	0.119	0.823	0.205
Management ->Effectiveness	0.066	0.069	0.085	0.780	0.218
Consumer Goods & Retail -> Efficiency	-0.113	-0.112	0.152	0.744	0.228
Sales -> Efficiency	-0.108	-0.094	0.148	0.733	0.232
Managing Director -> Efficiency	0.080	0.056	0.117	0.681	0.248
Negotiation Experience -> Efficiency	-0.111	-0.118	0.163	0.678	0.249
Automotive -> Efficiency	0.075	0.080	0.113	0.668	0.252
High-tech -> Efficiency	0.077	0.075	0.123	0.628	0.265
High-tech -> Effectiveness	0.065	0.066	0.123	0.529	0.299

Table 1.2. Complete List of Control Variables 2/2

	Original Sample	Sample Mean	Standard Deviation	T Statistics	P values
Organization Size -> Effectiveness	0.077	0.076	0.035	2.194	0.014
Divisional Head -> Effectiveness	-0.269	-0.270	0.126	2.133	0.016
Industrial Goods -> Effectiveness	0.262	0.254	0.123	2.132	0.017
Public -> Efficiency	-0.405	-0.397	0.195	2.072	0.019
Media & Telecommunications -> Efficiency	0.361	0.370	0.182	1.990	0.023
Team Lead -> Effectiveness	-0.198	-0.198	0.120	1.645	0.050
Media & Telecommunications -> Effectiveness	0.257	0.265	0.168	1.528	0.063
Chemicals -> Effectiveness	-0.175	-0.176	0.115	1.521	0.064
Management -> Efficiency	0.131	0.132	0.092	1.436	0.076
Negotiation History -> Effectiveness	0.094	0.094	0.069	1.361	0.087
Specialist -> Effectiveness	-0.173	-0.172	0.138	1.253	0.105
Purchasing -> Effectiveness	-0.106	-0.094	0.089	1.195	0.116
Industrial Goods -> Efficiency	-0.163	-0.176	0.143	1.140	0.127
Female -> Efficiency	0.084	0.075	0.077	1.078	0.140
Sales -> Effectiveness	-0.146	-0.149	0.139	1.051	0.147
Purchasing -> Efficiency	-0.103	-0.088	0.100	1.028	0.152
Department Head -> Efficiency	0.133	0.128	0.133	1.001	0.158
Services -> Effectiveness	0.096	0.102	0.100	0.964	0.167
Manager without Team -> Efficiency	0.121	0.117	0.126	0.962	0.168
Other -> Effectiveness	-0.253	-0.255	0.276	0.916	0.180
Manager without Team -> Effectiveness	-0.116	-0.104	0.135	0.856	0.196
Consumer Goods & Retail -> Effectiveness	-0.113	-0.114	0.132	0.856	0.196
Other -> Efficiency	-0.193	-0.183	0.226	0.853	0.197
Department Head -> Effectiveness	-0.098	-0.083	0.119	0.823	0.205
Management -> Effectiveness	0.066	0.069	0.085	0.780	0.218
Consumer Goods & Retail -> Efficiency	-0.113	-0.112	0.152	0.744	0.228
Sales -> Efficiency	-0.108	-0.094	0.148	0.733	0.232
Managing Director -> Efficiency	0.080	0.056	0.117	0.681	0.248
Negotiation Experience -> Efficiency	-0.111	-0.118	0.163	0.678	0.249
Automotive -> Efficiency	0.075	0.080	0.113	0.668	0.252
High-tech -> Efficiency	0.077	0.075	0.123	0.628	0.265
High-tech -> Effectiveness	0.065	0.066	0.123	0.529	0.299
Public -> Effectiveness	-0.077	-0.062	0.152	0.508	0.306
Negotiation Experience -> Effectiveness	0.068	0.060	0.164	0.416	0.339

Aus unserem Verlagsprogramm:

Michael Oryl
Support Software for Negotiation Preparation
An Interdisciplinary Research Approach
Hamburg 2024 / 224 Seiten / ISBN 978-3-339-13980-1

Josephine Schaumburg
Men are not better negotiators after all! Clarification and critical analysis of existing theory on gender and negotiation performance
Hamburg 2024 / 284 Seiten / ISBN 978-3-339-13798-2

Carsten Stork
Organizational Negotiation Management
Implementation and Impact
Hamburg 2023 / 190 Seiten / ISBN 978-3-339-13554-4

Niklas F. Bronnert
Introducing Automated Negotiators
Effects and Acceptance
Hamburg 2023 / 148 Seiten / ISBN 978-3-339-13514-8

Manuel Hefner
Choosing and Optimizing Negotiation Modes
Hamburg 2023 / 164 Seiten / ISBN 978-3-339-13376-2

Sandra Haggenmüller
Negotiation and its Changing Environment
Hamburg 2023 / 218 Seiten / ISBN 978-3-339-13366-3

Patricia Oehlschläger
Future Perspectives on Business Negotiations
Hamburg 2022 / 194 Seiten / ISBN 978-3-339-13256-7

Ernestine Cathérine Siebert
New Strategic Approaches for Multi-issue Negotiations
Hamburg 2022 / 198 Seiten / ISBN 978-3-339-12750-1

Andreas Wild
Strategic supplier cooperation in the automotive industry
Hamburg 2022 / 124 Seiten / ISBN 978-3-339-12618-4

Benjamin Hebisch
Power in supply chain negotiations: a two-stage approach
Hamburg 2021 / 140 Seiten / ISBN 978-3-339-12400-5

Iris Pöschl
Ambidextrous Negotiation Behavior
Conceptualization and Effects
Hamburg 2021 / 192 Seiten / ISBN 978-3-339-12284-1

VERLAG DR. KOVAČ
FACHVERLAG FÜR WISSENSCHAFTLICHE LITERATUR
Postfach 570142 · 22770 Hamburg · www.verlagdrkovac.de · info@verlagdrkovac.de